Lorin Low Dame, Frank Shipley Collins

Flora of Middlesex County, Massachusetts

Lorin Low Dame, Frank Shipley Collins

Flora of Middlesex County, Massachusetts

ISBN/EAN: 9783337268602

Printed in Europe, USA, Canada, Australia, Japan

Cover: Foto ©berggeist007 / pixelio.de

More available books at **www.hansebooks.com**

Middlesex County, Massachusetts,

——BY——

L. L. DAME AND F. S. COLLINS.

MALDEN:
MIDDLESEX INSTITUTE,
1888.

C. M. BARROWS & CO.,
PRINTERS,
2 NORTH MARKET STREET, BOSTON.

PREFACE.

This catalogue has been a natural outgrowth of the incorporation of the Middlesex Institute in 1881. The results of numerous excursions in the vicinity of Malden encouraged wider research, leading to the exploration of the entire county, with publication in view.

Works known to contain lists of county plants have been carefully examined, among which are Bigelow's "Florula Bostoniensis;" Hitchcock's "Report on the Geology, etc., of Mass.;" Emerson's "Trees and Shrubs of Mass.;" a "List of Plants in Malden and Medford;" the "Catalogue of the Davenport Herbarium;" "A partial list of the native flora of Waltham;" a "List of introduced plants found in the vicinity of a wool-scouring establishment" in North Chelmsford, by Rev. W. P. Alcott; and miscellaneous papers in the "Proceedings of the Boston Society of Natural History," Hovey's Magazine, and other publications.

The verification of these lists has been no light task. Where no specimen exists, where it has been impossible to find the original authority upon which a plant outside its recognized limits has been reported, or where there is a strong probability of wrong determination, the authors have excluded such names from the catalogue, thinking it better to lose several species which may be in the county than admit one which would be justly open to criticism. Various public and private collections have been examined for rare species; while holidays and vacations have been for six years largely devoted to botanical expeditions which have embraced within their scope, though very imperfectly, every section of Middlesex.

The catalogue follows Gray's Manual in the classification of the Phanerogams; the following exceptions to the order of the Manual are to be noted: species of Cyperus are arranged according to Dr. Britton's "A Preliminary List of North American Species of Cyperus, with Descriptions of New Forms." Bull. Torr. Bot. Club, Vol. XIII., No. II.; species of Carex following Prof. Bailey's "A Preliminary Synopsis of North American Carices," etc., contributed April 14, 1886 to the Proceedings of the Amer. Acad. of Arts and Sciences; genera of Grasses after Bentham and Hooker's Genera Plantarum.

The Cryptogams follow the arrangement suggested in Gray's "Lessons in Botany," revised edition, 1887. The nomenclature follows Gray in the North Am. Flora, the Revision of the Ranunculaceæ, and of Violaceæ; Coulter in the Umbelliferæ; Watson's Index elsewhere in the Polypetalæ; and Morong in the Typhaceæ. In the Cryptogams, the catalogue adopts the classification of Engelmann on Isoetes, James and Lesquereux on Mosses, and Farlow on marine Algæ.

Many botanists have kindly contributed their services. Dr. Gray determined the doubtful species of Compositæ, more especially the Golden Rods and Asters. Dr. Sereno Watson has given advice and information, and afforded special facilities for work at the Gray Herbarium. Dr. W. G. Farlow has rendered assistance in the Characeæ and fresh water Algæ. Mr. M. S. Bebb has gone over specimens of our county willows, and contributed notes upon several species. Rev. Thos. Morong has furnished notes upon the Naiadaceæ, and a set of Potamogetons for the county herbarium. Miss Clara E. Cummings has revised the list of Mosses and Lichens, and Mr. Geo. E. Davenport the ferns. Dr. D. F. Lincoln has given an account of the geological characteristics of the soil.

Many of the earlier determinations in the Juncaceæ and Cyperaceæ were made by Mr. Wm. Boott; the later, in the genus Cyperus, by Dr. N. L. Britton, of Columbia College, New York; in the genus Carex, by Prof. L. H. Bailey, Jr., of the Agricultural College at Lansing, Michigan; in the Gramineæ, from time to time, by Dr. Geo. Vasey and Prof. F. Lamson Scribner, of the U. S. Department of Agriculture, Washington, D. C.

Thanks are specially due to Dr. C. W. Swan, of Boston, who has put at the disposal of the authors his valuable herbarium, his extensive acquaintance with the county flora, and his personal services from the beginning to the end of the flora. Without his critical labors upon the Carices and Gramineæ, the catalogue could hardly have been ready the present season. Thanks are due likewise to W. H. Manning, Walter Deane, C. E. Faxon, Edwin Faxon, Mrs. S. E. French, E. S. Hoar, Mrs. P. D. Richards, Dr. F. Nickerson, C. W. Jenks, H. A. Young, William Edwards, Rev. J. H. Temple, Miss A. M. Symmes, Miss Emily F. Fletcher, and others whose names appear in the following pages.

The catalogue does not claim to be exhaustive; while the Phanerogams, Vascular Cryptogams and marine Algæ are as complete, perhaps, as may be expected in any list covering so much ground, many additions will undoubtedly be made, more especially in the north-western sections. The remaining Cryptogams are simply a contribution for the benefit of special students.

PLAN OF CATALOGUE.

1. Names of plants thought to be indigenous have been printed in heavy, broad-face type.
2. Names of introduced plants, propagating freely by

seed or suckers beyond the limits of cultivated ground, and likely to survive, unless destroyed by the agency of man, are printed in small capitals.

3. Italics have been reserved for introduced species which have not become permanently established within our limits. These adventive species range from plants which have been found but once, like some of the wool-waste Compositæ of Lowell and Chelmsford, to familiar species, for example, Lucerne, which persists for many years, but does not appear to thrive, and in a few cases observed through a period of several years, has died out altogether. Within this division, also, are included many plants, such as the wool-waste Medicks, which spring up in abundance every year; it is not settled, however, whether they are perpetuated by seed ripened in Middlesex and surviving the winter, or by fresh importations from the original source.

4. To facilitate the study of species not contained in the Manual, either a description has been given, or reference made to some volume of the Wood or Gray series of text-books wherein such description may be found.

5. Whenever a plant is commonly met with in its proper habitat, no location is given. Where several stations are given without comment, the species will probably be found more widely distributed. The occasional presence of a plant in stations besides those mentioned is indicated by the abbreviation *et al.*

6. Wherever the "Manual" is mentioned, reference is made to Gray's Manual, 5th edition, unless otherwise designated.

7. In the Phanerogams and Vascular Cryptogams, an asterisk indicates that no specimen of the plant so designated is in the county herbarium. It does not follow, however, that such species are always rare.

INTRODUCTION.

Middlesex county is very irregular in outline, comprising an area, roughly approximated, of 830 square miles. On the north it borders upon the New Hampshire line, an extreme point in Dracut reaching to latitude 42° 44' 12", on the east it extends (in Malden) to longitude 71° 54', closely approaching the sea coast; on the south, it touches lat. 42° 9' 30" (Holliston); and extends westward in a single tier of towns to long. 71° 1' 30" (Ashby).

The highly diversified character of the county may best be seen by a glance at the accompanying map and key. The Merrimac river passes through the north-eastern section, and the Concord through the centre, while all portions are abundantly watered by numerous smaller rivers and creeks. There are one hundred and thirty-seven ponds, some of them of considerable size; numerous swamps of greater or less extent; and salt marshes along the tidal streams. The surface is very uneven, Prospect Hill, Waltham, reaching a height of 482 ft.; Reservoir Hill, Lincoln, 395 ft.; Goodman Hill, Sudbury, 415 ft.; Reeves Hill, Wayland, 410 ft.; Pegan Hill, Natick, 408 ft.; Nobscot Hill, Framingham, 602 ft. The general elevation gradually rises to the highlands of Townsend and Ashby, culminating in Mt. Watatic, a granitic mass, 1847 ft. above the sea level, the highest land in the county.

The geological ages represented in Middlesex county are three, viz.:

1. The Cambrian, with slates and conglomerate; it includes the comparatively low land within four to six miles of Boston.

2. The Huronian, just outside of this, beginning at and including the Middlesex Fells, comprises primitive rocks—granite, petrosilex (porphyry), diorite, hornblendic gneiss, quartzite, quartzy slate. Limestone occurs in patches.

3. The Montalban (later in date than the Huronian, but antecedent to the Cambrian), lies to the north and west of the Huronian, and comprises granite, gneiss, mica slate, argillite, and numerous patches of limestone. It occupies the larger part of the county, and is bounded by a N. E. to S. W. line which includes the towns of Wilmington, Bedford, Concord, Sudbury, and Framingham.

The soil, like that of most parts of New England, is mainly dependent for its characteristics upon the glacial drift, which covers most of the rocks to the depth of many feet. This material consists of two portions; the very compact boulder clay or till, often called "hard pan;" and a loose mass, of gravelly and sandy consistency, which has been derived from the boulder clay by the the washing of ancient torrents. The latter is often stratified, is comparatively free from boulders, and forms the present soil, with such additions as the yearly decay of vegetation for many centuries has made. Its qualities vary greatly, some having been deposited in the form of sand, or sterile gravel, while other parts are of a rich, clayey nature. "Terraces of sand and gravel from the re-assorted boulder clay make up by far the greater part of the low-lying, arable lands of eastern Massachusetts; and of this nature are about all the lands first used for town-sites and tillage by the colonists, notwithstanding the soil they afford is not as rich or as enduring as the soils upon the unchanged boulder clay." N. S. Shaler, Memorial History of Boston, vol. 1, p. 6.

As a modifying agent, it is necessary to keep in mind the many small areas of lime rock, which lie so scattered as to make enumeration difficult. Details are given in Crosby's map of eastern Massachusetts.

Opinions may differ as to the precise mode of origin of the materials composing the drift, but there can be no doubt of the main fact that they represent the rocks with which we are familiar in New England, mostly consisting of mica schist, gneiss, and the like. In these rocks feldspars abound, containing much potash, soda and lime,— materials which become of use when decay has reduced the rock to the condition of mud or clay.

It is not probable that the decay of these rocks *in situ* has contributed much to the soil of the region. They are mostly durable, and it is exceptional to find them decayed to any, great depth. The period of time, too, during which such decay is conceivable, is a relatively short one — since modern observers incline to the view that only from 6000 to 10,000 years have elapsed since the glacial period.

The diversity of physical conditions gives rise to a corresponding variety in the character of the flora. **Potentilla tridentata, Vaccinium Canadense, Ribes prostratum, Acer spicatum, Abies balsamea, Taxus baccata,** var. **Canadensis,** and **Dalibarda repens** have been found indisputably native only upon or near Mt. Watatic; **Viola rotundifolia** in a deep ravine at Ashby; **Dirca palustris** and **Lonicera cœrulea,** at Townsend; Alpine lichens and mosses are occasional upon high hills; while **Ledum latifolium, Kalmia glauca, Andromedia polifolia, Chiogenes** and **Smilacina trifolia** linger here and there in cold sphagnum swamps.

A few more southern species are sometimes met with, among which are **Draba Caroliniana,** Woburn; **Draba** verna, Medford; **Smilax glauca,** Weston; the very rare **Habenaria ciliaris,** Lexington; **Pycnanthemum linifolium,** Reading; **Asclepias verticillata,** in several localities.

Perhaps the most peculiar flora occupies the tract lying between Horn Pond Mountain and Winchester, including Winter Pond and a small sheet of water nearly dry in midsummer, to which the name Round Pond has been given by botanists. About the borders of both these ponds grows **Coreopsis rosea** in abundance; and on the shores of the former **Ludwigia polycarpa, Eleocharis Engelmanni,** var. **detonsa, Scleria reticularis,** and **Cuscuta arvensis.** It is difficult to frame a satisfactory theory for the presence of this little colony.

Great care has been taken to mark clearly the distinction between species believed to be indigenous within the county limits, and those introduced from without the county, whether from the old world, remote sections of America, or even other parts of Massachusetts. To exclude naturalized species would be to exclude some of our most common plants, and a total of fully one-sixth of our Phanerogams. The difficulty lies in drawing the line between the lately or locally naturalized and the purely adventive. Many introduced plants, now occupying very limited areas, will surely abide with us, if undisturbed by man. They run no greater risk of extermination than many of our attractive native plants.

There seems to be abundant reason for cataloguing adventive plants, provided their status be appropriately indicated. They are candidates for naturalization. Indeed, when a plant is called indigenous, the term implies simply that, as far back as any record exists, it was a part of the flora. New plants have always been creeping in: water-courses, winds and birds of passage are constantly spreading the area of special forms of plant life. When the seed finds a suitable environment, it develops, and the plant multiplies oftentimes with astonishing rapidity. Within a hundred years every trace of its foreign origin may disappear.

Very many species, too, have been introduced through the indirect agency of man. The highways and byways mark the line of march of an invading army. Middlesex county has also long been a manufacturing centre, and about the cotton and more especially the woollen mills, a strange flora is striving to adapt itself to new climatic conditions. Some of these immigrants have undoubtedly come to stay.

Other plants, directly introduced, have become thoroughly established, among which may be mentioned the Privet, now common everywhere about Boston. The late Minot Pratt, an enthusiastic botanist, throughout a period of forty years sought to naturalize within the limits of Concord plants from all sections of the United States. Some few of these have disappeared altogether, others maintain a precarious existence, while still others have abundantly increased, in some cases even becoming troublesome weeds. As these plants were skilfully set out in situations to correspond with their natural habitat, they have often been found by occasional collectors, and reported as indigenous. For this reason it has been thought best to incorporate in the present work a complete list of such plants, taken from a manuscript volume left by Mr. Pratt to the Concord Public Library.

Middlesex county has been fortunate in the location within its limits of the Harvard Botanic Garden, the headquarters of a corps of able naturalists, and a centre of scientific activity. It has also been fortunate in affording to the botanists of the neighboring city of Boston a convenient field for exploration.

In 1814, Dr. Jacob Bigelow published his Florula Bostoniensis, which embraced in its plan a considerable portion of Middlesex. This work, which passed through a second edition in 1824, and a third in 1840, became at once a standard authority, and gave a mighty impulse to the study of botany.

For more than a quarter of a century, B. D. Greene, a keen and accurate observer, whose ample fortune happily left him at liberty to pursue his favorite study, herborized extensively in Tewksbury and adjacent towns. He was also an unconscious contributor to the Desmid flora of the county. On specimens of Utricularia collected by him at Tewksbury and transmitted to Swedish herbaria, Lagerheim has detected seven new species and fourteen new varieties of Desmids which are enumerated in their place among the Algæ.

Prof. Edward Tuckerman, while residing in Boston or vicinity, contributed largely to our knowledge of the county flora, more especially of the Lichens, of which he made an extensive collection.

Rev. J. L. Russell, a diligent student of Cryptogamic botany, while settled at Chelmsford, made collections of Musci, Hepaticæ and Lichens, publishing from time to time the result of his researches.

George B. Emerson, 1840-1845, repeatedly traversed Middlesex, as indeed every other county of the state, in search of material for his "Report on the Trees and Shrubs growing naturally in the Forests of Massachusetts," which appeared in 1846, and has been used ever since as the best available text-book for the study of our trees.

Charles E. Perkins, whose early death in 1883 cut short a botanical career of much promise, had been at work for several years gathering data for a Flora of Boston and vicinity. His notes and collections, bequeathed to the Middlesex Institute, have been freely drawn upon in the present catalogue.

Of the local botanists, however, who have passed on, we are most indebted to Wm. Boott, who was born in Boston in 1805, and died in the same city in 1887. "His tastes and accomplishments in early and middle life" writes Dr. Gray, " were literary, especially linguistic. Probably he took up botany at the instigation of his brother (Dr. Francis Boott, of London), and with the design of helping him to the Carices of this country, when Dr. Boott began the study of this vast genus of which he became the illustrator and highest authority ; and Wm. Boott, by a kind of *noblesse oblige*, after his brother's death, devoted himself to this study." He likewise studied critically the Potamogetons, Isoetes, the Grasses and some tribes of the Cyperaceæ. He was, moreover, a good general botanist, with whom the zeal of the collector and the uneasy spirit of original research abode to the last. A rare combination of painstaking care and critical acumen made his determinations authoritative, while the many summers he spent in Medford gave him an extraordinary acquaintance with the flora of the neighborhood. To the preparation of this work, he contributed a list of Middlesex plants, specimens from his herbarium, and his personal services in the identification of doubtful species. Wm. Boott has left in print a scanty record, but his herbarium, bequeathed to Harvard University, gives a partial idea of the scope of his labors.

The number of trained observers now in the field is a guarantee that the work of their predecessors will be worthily continued. Discoveries in every class of plants may confidently be expected.

The annexed species and varieties have been founded on specimens first collected in Middlesex county.

Rubus setosus. Bigelow, Fl. Bost., 2 ed., p. 98.
"In a swamp at Sudbury." Since reduced to a variety, Rubus hispidus, L., var. setosus, Torr. & Gray. Fl. 1, 456.

Myrophyllum tenellum. Bigelow, Fl. Bost., 2 ed., p. 346.
"Edge of Fresh Pond, and also at Tewksbury."

Utricularia resupinata. B. D. Greene in Bigelow's Fl. Bost., 3 ed., p. 10.

Potamogeton gramineus, L., var, **spathulæformis,** Robbins.
Based upon P. spathæformis, in herb. Tuckerman. "Mystic Pond, near Boston." Man., p. 487.

Potamogeton gramineus, L., var. **maximus.** Morong in Middlesex Fl., p. 100.

Potamogeton Mysticus. Morong in Bot. Gaz., Vol. V., No. 5.

Potamogeton pusillus, L., var. **gemmiparus,** Robbins.
Based upon P. gemmiparus, in herb. Robbins. "Outlet of Mystic Pond, near Boston." Man., p. 489. Now restored by Morong to its former specific rank and name, **P. gemmiparus,** Robbins.

Naias flexilis, Rostk., var. **robusta.** Morong in Bot. Gaz., Vol. X., No. 4.
"Concord river."

Juncus militaris. Bigelow, Fl. Bost., 2 ed., p. 39.
"In a pond at Tewksbury."

Asplenium ebeneum, Ait., var. **serratum,** Gray.
"Malden, Nov., 1872." G. E. Davenport, in Cat. of the Davenport Herb.

Aspidium Boottii. Tuckerman, in Hovey's Mag., Vol. IX., p. 145 (1843). (A. spinulosum, Swartz, var. Boottii, Man.)
Lowell (Wm. Boott).

Isoetes Tuckermani, Braun.
Mystic River and Pond, 1848 (E. Tuckerman).

Isoetes echinospora, Durieu, var. **Boottii.** Engelman in Man., p. 676.
Based upon Isoetes Boottii, Braun *in litt.* "Pond in Woburn, near Boston, 1867, partly out of water." (Wm. Boott).

Isoetes echinospora, Durieu, var. **muricata,** Engelman, in Man., p. 676.
"Woburn creek and Abajona river" (Wm. Boott).

To the above list may be added seven new species and fourteen varieties of Desmids. (See p. 159).

ABBREVIATIONS.

A. Br., Monogr.—A. Braun, Monographie der Characeen.

Adv.—Adventive.

Am. Nat.—American Naturalist.

Bailey, Prel. Syn. N. A. Carices.— Preliminary Synopsis of North American Carices.

B. S. N. H.—Boston Society of Natural History.

Bigelow's Fl. Bost.—Jacob Bigelow's Florula Bostoniensis.

Boiss. Fl. Or.—Boissier, Flora Orientalis.

Bot. Cal.—Botany of California, Watson.

Bot. Reg.—Botanical Register.

Chapman's S. Fl.—Flora of the Southern States.

Coulter, R. M. Bot.—Manual of the Botany of the Rocky Mountain Regions.

DC., Prodr.—De Candolle's Prodromus.

Gray, Syn. Fl. N. A.—Synoptical Flora of North America.

Int.—Introduced.

Koch, Syn. Flor. Germ.—Synopsis Floræ Germanicæ.

Koch., Taschenb. d. Deutsch. & Schw. Fl.—Taschenbuch der Deutschen und Schweizerischen Flora.

Lesq. & James, Man.—Lesquereux and James, Manual of the Mosses of North America.

Man.—Gray's Man., Fifth Edition.

Nat.—Naturalized.

Wood's Bot. & Fl.—Botanist and Florist.

A

TOPOGRAPHICAL

REFERENCE MAP

— OF —

MIDDLESEX COUNTY,

MASSACHUSETTS,

WITH KEY,

Showing the location of Villages, Hills, Ponds, Brooks, Swamps, etc.,
Compiled from the latest revised Maps

— BY —

EDWARD P. ADAMS,

1885.

MAP OF MIDDLESEX COUNTY.

In the scientific study, in any branch of natural history, of a particular section of country, and especially in the study of its flora, constant reference must be made to localities that can be described only by some topographical feature. It is a convenience, if not a necessity, to know the relative position of these features. It was to supply this want that the present map was compiled. By using reference letters and the key instead of the names in full upon the map, as is usual, a map of pocket size was made to show more topographical features by name than have been shown upon any wall map of the county.

Middlesex Fells, comprising about four thousand acres, extends from Pine Hill in Medford, on the south, to Bear Hill, in Stoneham, on the north; and from Winchester easterly as far as the B. & M. R. R. in Malden and Melrose. No other large tract of natural growth in the county has a distinctive name, although some smaller places, such as Shaker Glen in the western part of Lexington, and Pine Banks between Malden and Melrose, have received special names from those who frequent them.

E. P. A.

MEDFORD, Mass., May 4, 1888.

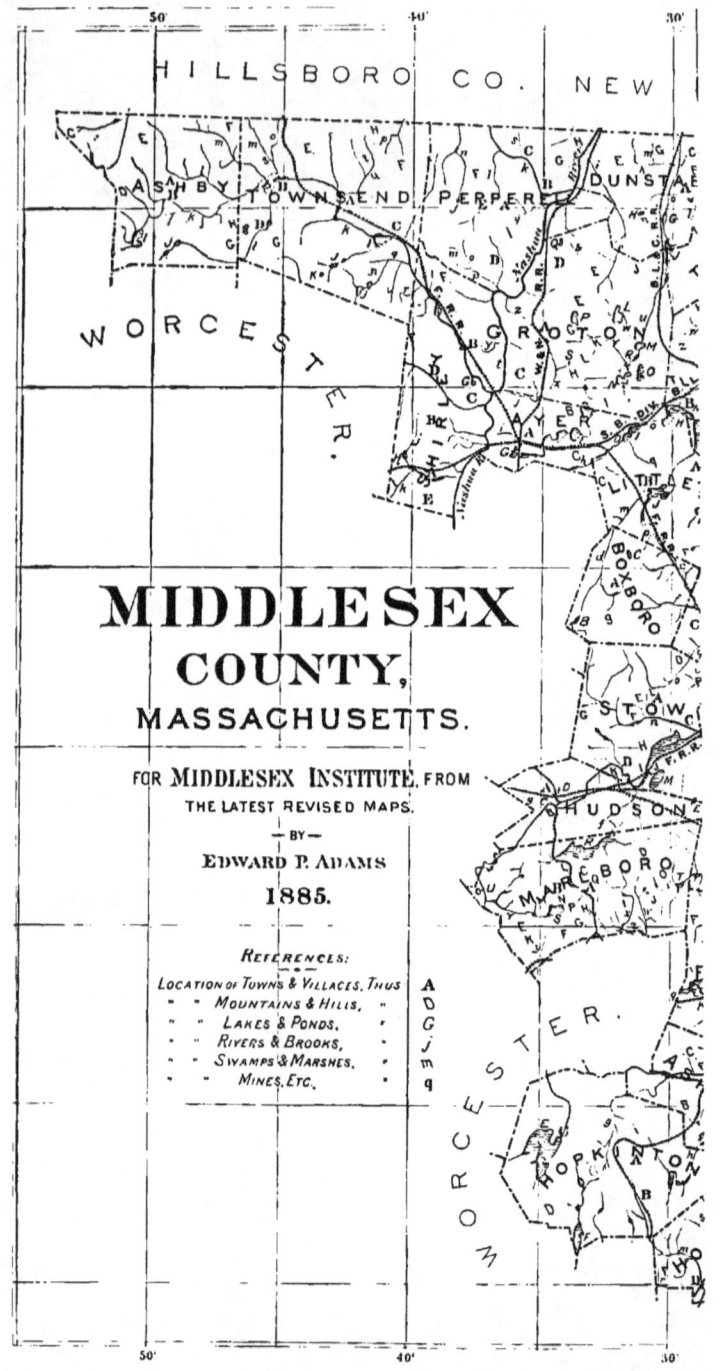

MIDDLESEX
COUNTY,
MASSACHUSETTS.

FOR MIDDLESEX INSTITUTE, FROM
THE LATEST REVISED MAPS.

— BY —

EDWARD P. ADAMS

1885.

REFERENCES:

LOCATION OF TOWNS & VILLAGES, THUS A
" " MOUNTAINS & HILLS, " D
" " LAKES & PONDS, " G
" " RIVERS & BROOKS, " J
" " SWAMPS & MARSHES, " e
" " MINES, ETC. " q

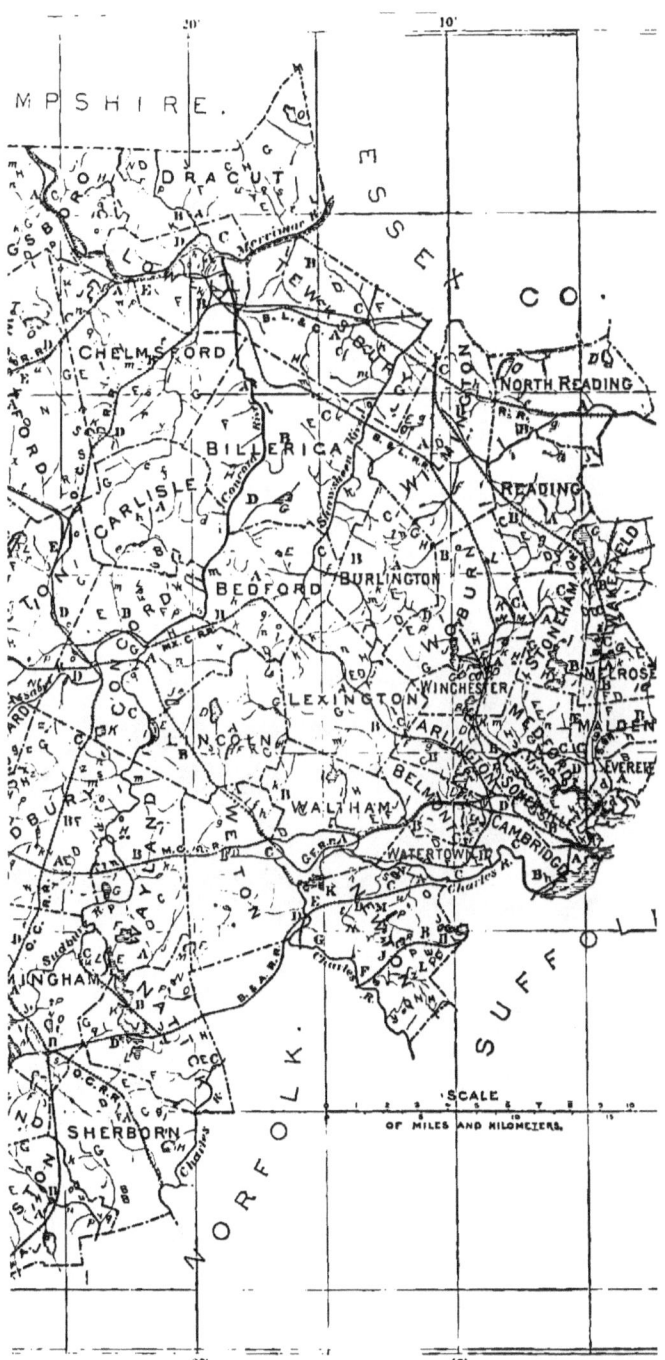

Towns.	Villages.	Hills.	Ponds.	Brooks.	Swamps.
Acton	A Acton P. O. B S. Acton C W. Acton D E. Acton E N. Acton	F Nagog G Strawberry H Great I Hearthstone J Myer's K Conant	L Nagog M Grassy N Mill	o Nashoba p Fort Pond q Heathen Meadow r Nagog	s South Acton
Arlington	A Arlington B Arlington Heights	C Arlington Heights	D Mystic E Spy F Arlington Reservoir	g Vine h Alewife	
Ashby	A Ashby Centre B Mill Village	C Mt. Watatic 1847 ft. D Mt. Pleasant 1597 ft. E Jones's F Juniper G Rattlesnake H Fort	I Reservoir J Wright's or Nocsepogesuck	k Willard l Locke m Trap Falls	
Ashland	A Ashland	B Magunco C Wild Cat D Ballard's E Banner	F Waushakum G Cold Spring H Reservoir	i Cold Spring	j Guinea Meadow
Ayer	A Ayer Junction	B Snake C Brown Loaf	D Spectacle E Sandy F Long G Plow Shop	h Bennett's i Nonacanicus j James	
Bedford	A Bedford Centre B W. Bedford	C Indian D Silver	E Fawn Lake	f Vine g Hartwell h Elm i Spring j Peppergrass k Trout l Kiln	m Bedford Meadow n Yellow Ochre Bed o Marl and Clay Bed
Belmont	A Belmont B Waverly C Mt. Auburn		D Little	e Little River f Beaver	
Billerica	A N. Billerica B Billerica C Billerica Station D S. Billerica	E Fox F Gilson's	G Nutting H Winning's	i Wright j Webb's	k Shawsheen river l Concord river m Swamp
Boxborough	A Boxborough		B Withington C Muddy	d Beaver e Half Moon f Guggin's	g Wolf
Burlington	A Burlington Centre B Hovenville	C Wood D Babylon E Greenleaf Mount and "Garden of Eden" Rav.	G Sawmill H Tannery	i Vine j Long Meadow k Sandy l Ipswich River	m Long Meadow n Cranberry Meadow

TOWNS.	VILLAGES.	HILLS.	PONDS.	BROOKS.	SWAMPS.
Cambridge	A E. Cambridge B Cambridgeport C Old Cambridge D N. Cambridge		E Fresh	g Alewife	h Salt marsh
Carlisle	A Carlisle Centre	B Bellows G Rail Tree		c River Meadow d Page's e Spencer	f Tophet h Copper Mine i Iron Ore
Chelmsford	A N. Chelmsford B Chelmsford C W. Chelmsford D S. Chelmsford	F Francis G Robin's H Rocky I Pine I Chestnut	J Newfield or Amannicumsick K Baptist or Hart	l River Meadow m Beaver n Stony o Deep p Black q Crooked Spring r Farley s Putnam t Golden Cove u Scottie v House	w Black Brook x River Meadow
Concord	A Concord B Plain C Nine Acre Corner D Damon's Mills or Westvale	E Ampenseack F Ponkatasset G Lee's	I Fair Haven Bay J Walden K White L Bateman's	m Spencer n Mill o Nashoba p Saw Mill	q Sudbury River r Assabet River s Concord River t Mill Brook u Bateman's Pond v Hill w Saw Mill Brook
Dracut	A Dracut P. O. or "The Parish" B Merrimac Mills	C Marsh D Chandler E Loon F Bump G Burns H Thornton I Whortleberry J Winter K Ledge L Mine Pit M Poplar	N Long O Peters	p Beaver River q Richardson's r Donble s Trout t Varnum's u Potash	v Richardson's Brook
Dunstable	A Dunstable P. O.	B Roby C Kendall E Hound Meadow F Nutting G Blanchard's	G Lower Massapoag H Mill	i Unkety, or Unemenassette j Salmon k Joint Grass l Barnes'	n Mill Pond o Salmon Brook
Everett	A Everett	B Belmont C Mt. Washington			d Salt Marsh
Framingham	A Framingham Centre B S. Framingham	E The Mountain F Gibbs Mountain	K Boston Water Works Reservoirs	q Stony Brook r Baiting	v Guinea Meadow

TOWNS.	VILLAGES.	HILLS.	PONDS.	BROOKS.	SWAMPS.
Framingham	C Saxonville D Nobscot, or N. Framingham	G Nobscot H Merriam's, or Pinca-shion I Mt. Wake J Bare	L Lake Cochituate M Farm N Learned's O Gleason's	s Beaver Dam t Hop u Cochituate	r Cow Pond Meadow
Groton	A Groton B W. Groton	E Chestnut F The Throne G Gibbet H Indian I Rocky J Horse K Brown Loaf L Prescott	L Baddacook M Cow N Knop's O Duck P Martin's Q Wattle's S Cady	t Nashua River u Cow Pond v Martin's Pond w Baddacook x James y Wrangling z Gratuity $ Cold Spring	
Holliston	A Holliston B E. Holliston C Metcalf's D Bragg'ville	E Bald F Rocky Woods G Long H Mt. Hollis I Powder House	J Winthrop	k Beaver Dam l Chicken m Hopping n Jar o Dapping p Bogastere y Dirty Meadow r Deer	s Cedar t Burnt u Dapping Brook v Bogastere w Hopping Brook
Hopkinton	A Hopkinton Centre B Hayden Row C Woodville		F Whitehall F North	g Indian h Cold Spring	i Cedar
Hudson	A Hudson P. O.	B Mt. Assabet	C Mill D Mill E White	f Meadow g Hog	
Lexington	A Lexington Village B E. Lexington	C Mt. Ephraim D Granny's E Merriam's F Turner's G Loring H Buck's	I Arlington Reservoir	j Vine	k Tophet m Bryant n Chandler
Lincoln	A Lincoln B S. Lincoln	C Cave	D Sandy E Fairhaven F Beaver G Ice H Mill		
Littleton	A Littleton P. O. B W. Littleton	C Oak D Long Pond or Brothers' E Nashoba F Newtown	G Nagog H Forge I Spectacle J Mill K Fort L Long	m Beaver n Long Pond o Stony p Half Moon	q Brown's
Lowell	A Lowell B Ayer City	F Lowell Highland		q Pawtucket Canal k River Meadow	

TOWNS.	VILLAGES.	HILLS.	PONDS.	BROOKS.	SWAMPS.
Lowell	C Centralville D Pawtucketville E Middlesex Village			*h* Northern Canal *i* Western Canal *j* Middlesex Canal	
Malden	A Malden Centre B Maplewood C Edgeworth D Linden E Malden Highlands	F Wait's Mount.		*g* Malden River	*h* Rich's Meadow
Marlborough	A Marlborough P. O. B West Village	C Ockooanganset D Spoon E Stirrup F Jericho G Shoestring H Maynard's or Fairmount I Farm Hill J Indian Head K Crane L Addition M Bear N Slygo O Mt. Ward P Pleasant Q Prospect	R Fort Meadow Reservoir Q William's or Gate's R Saw Mill U Muddy	*v* Milburn *w* Broad Meadow *x* Mowry *y* Road	*z* Broad Meadow
Maynard	A. Maynard P. O.	B Pompasiticut, or Summer C Silver D Vose's	E Puffer's F Vose's		*g* Meadow
Medford	A Medford B W. Medford C Glenwood D Wellington	E Pine F Mt. Andrew G Pasture H Flagstaff I College J Rock	K Mystic	*l* Smelt *m* Brooks' *n* Floyd's	*o* Salt Marsh
Melrose	A Melrose B Wyoming C Melrose Highland	D High Rock E Pine F Pine Banks	G Crystal Lake H Long I Swam's	*j* Spot Pond *k* Lynde	
Natick	A Natick B Felchville C S. Natick D Walkerville	E Carver's or Badger's F Train's G Tom's H Broad's I Walnut J Steep Rock	K Lake Cochituate L Dug M Nonesuch N Pickerel O Jennings' P Mud	*q* Steep	
Newton	A Newton P. O. B Newton Centre C Newtonville	J Waban K Moffatt L Chestnut	O Chestnut H. Reservoir P Bullough's Q Hammond's	*t* Cheesecake *u* Laundry *v* Cold Spring	*w* Hammond's Pond *x* Cold Spring *y* Nahanton St.

Towns.	Villages.	Hills.	Ponds.	Brooks.	Swamps.
Newton	D W. Newton E Auburndale F Newton Upper Falls G Newton Lower Falls H Chestnut Hill J Newton Highlands L Thompsonville N North Village	M Bald Pate N Oak O Mt. Ida P Institution Q Nonantum	R Crystal Lake S Silver Lake		z Parker St.
No. Reading	A N. Reading B Pudding Point		C Martin's D Swan	e Martin's f Raper's g Ipswich River	
Pepperell	A Pepperell P. O. B E. Pepperell C North Village D S. Pepperell E Burkinshaw's Factory	F Oak G Nissitissitt H Pine I Clarke's	J Heald's	k Nissitissitt River l Sucker m Robinson's n Blood's or Gulf o Bancroft p Nutting q Greene's r Varnum's s Mine	
Reading	A Reading P. O. B Reading Highlands	C Long Woods D Bear E Scotland		g Saugus River h Bear Meadow	i Hundred acre Mead. j Cedar
Sherborn	A Sherborn P. O.	B Nason or Pocasset C Pine D Brush E Peter's F Paul G Dirty Meadow	H Farm Lake I Little	j Dirty Meadow k Dopping or Dopple	l Dirty Meadow
Shirley	A Shirley Village B Shirley Centre C Woodsville D N. Shirley E Shaker Village	F Hunting	G Squannacook H Dead		l Tophet
Somerville.	A E. Somerville B Union Square C W. Somerville	D Central E Winter F Prospect G Spring H Convent or Mt. Benedict		i Alewife	j Mystic River Marsh k Milldam Marsh
Stoneham.	A Stoneham B Haywardville C Farm Hill	D Greenwood Mt. E Taylor Mt. F Mill Lot G Bear	H Spot I Doleful	j Spot Pond	k Turkey
Stow.	A Stow Centre	D Flagg	M Boon's	n Assabet	p Heathen Meadow

Towns.	Villages.	Hills.	Ponds.	Brooks.	Swamps.
Stow	B Rock Bottom C Lower Village	E Marble F Spindle G Long H Birch I Honeypot J Boon's K Orchard L Warren		o Heathen Meadow	
Sudbury	A S. Sudbury or Mill Village B Sudbury C N. Sudbury	D Goodman's E Green F Pendleton's G Flagstaff H Willis I Round	H Willis I Blandford J Bottomless	k Wash l Hop m Pantry n Cold o New Bridge p Run q Dudley r Laudham	s Gulf Meadows t Cold Brook Mead. u Bottomless Pond " v Dudley Meadows. w Laudham Brook " y Sudbury Meadows. x Iron Ore Bog.
Tewksbury	A Tewksbury B N. Tewksbury C Tewksbury Junction	D Prospect E Strongwater or Job's F Pinnacle G Snake	H Long I Round J Mud	k Strongwater l Trull m Heath	n Great Swamp. o Shawsheen River
Townsend	A Townsend P. O. B W. Townsend C Townsend Harbor D Rogerville	E Barker F Townsend G Raberry H West	I Harbor J Reservoirs K Dead	k Squannacook River l Pearl Hill m Walker n Witch o Mason p Wolf q Trout r Punkin	s Ash t Dead u Wolf
Tyngsbro'gh	A Tyngsborough	B Scribner's C Perham D Oak E Pine F Spectacle Hill and Fox Ledge G Abraham's H Locust	H Tyng's I Mud, or Park J Massapoag Lake	j Lawrence k Bridge Meadow l Waldo m Howard's n Blodget's o Scarlet	p Tyng's q Butterfield
Wakefield	A Wakefield B Wakefield Junction C Greenwood	D Cowdry E Greenwood Mt. F Round	G Lake Quannapowitt H Crystal Lake		
Waltham	A Waltham B Prospectville	C Prospect D Bear E Cedar F Owl G Reservoir	I Mean's or Hardy's	j Clematis k Hobb's l Beaver	
Watertown	A Watertown B Mt. Auburn C U. S. Arsenal	D Meeting House E Coolidge			

Towns.	Villages.	Hills.	Ponds.	Brooks.	Swamps.
Wayland	A Cochituate B Wayland Centre	C Reeve's D Overdrow	E Long Pond or Lake Cochituate F Dudley G Pelham's or Heard's H Baldwin's I Folsom	j Snake k Hayward's l Landham m Trout	n Sweetham Meadow o Sedge " p Beaver Hole "
Westford	A Westford P. O. B Forge Village C Graniteville D Brookside Sta. E Westford Sta.	F Westford G Francis H Snake Meadow I Oak J Flushing K Conscience L Kiascook M Prospect N Providence O Blake's P Burns' Q Spark's R Bear S Little Bear T Clay Pit	M Forge N Beaver Brook Flowed meadow O Nabnasset P Long-sought-for Q Keyes' R Burgess S Hart T Flushing	t Nashoba u Stony v Keyes w Meadow x Noneset y Gilson's	z Snake
Weston	A N. Village B Weston Centre C Stony Brook D Riverside		E Nonesuch	f Stony g Cherry h Hobbs'	
Wilmington	A Wilmington B N. Wilmington C Wilmington Junction	D Pine	E Silver Lake	f Maple Meadow g Lubber h Martin's	
Winchester	A Winchester P. O.		B Mystic C Wedge D Judkins E Abajona F Cutter's G Winter H High Serv. Reserv'r	i Abajona River j Horn Pond k Abajona	
Woburn	A Woburn Centre B N. Woburn C E. Woburn D Cummingsville	E Whispering F Horn Pond Mt. G Zion's H Mt. Pleasant I Rag Rock	J Horn K Burbank L Richardson's M Hill or Woodrough N Paddock or Whitemore	o Willow p Horn Pond	

CATALOGUE OF PLANTS.

PHÆNOGAMIA.

EXOGENS.

RANUNCULACEÆ. CROWFOOT FAMILY.

CLEMATIS, L.

C. Virginiana, L. VIRGIN'S BOWER. CLEMATIS.
Common. July-Aug.

ANEMONE, L.

A. cylindrica, Gray. LONG-FRUITED ANEMONE.
Westford (Misses Fletcher and Hodgman); Holliston (F. S. Collins); Townsend (Miss H. E. Haynes); Medford (Wm. Boott). Not common. May-June.

A. Virginiana, L.
Rather common. June-July.

A. nemorosa, L. WOOD ANEMONE. WIND-FLOWER.
Common. April-May.

A. Hepatica, L. (Hepatica triloba, Chaix; Man.) HEPATICA.
Rather common. April-May.

*A. acutiloba, Lawson. Hepatica acutiloba, DC.; Man.)
Concord; introduced from Vermont by Minot Pratt. April-May.

ANEMONELLA, Spach.

A. thalictroides, Spach. (Thalictrum anemonoides, Michx.; Man.) RUE ANEMONE.
Common. April-June.

THALICTRUM, Tourn.

T. dioicum, L. EARLY MEADOW-RUE.
Not rare. April-May.

T. purpurascens, L. PURPLISH MEADOW-RUE.
Malden (R. Frohock); Medford and Waltham (Wm. Boott); Townsend (Miss H. E. Haynes); Woburn (C. E. Perkins); Lowell (Dr. C. W. Swan). The form in this county seems to be var. **ceriferum** of C. F. Austin. Not common. June.

T. polygamum, Muhl. (T. Cornuti, L.; Man.) TALL MEADOW-RUE.
Very common. June-Sept.

RANUNCULUS, L.

*R. aquatilis, L. (R. aquatilis, L., var. heterophyllus, DC.; Man.);
FLOATING WATER-CROWFOOT.
Newton (Bigelow's Fl. Bost.) "Not met with for many years;
was possibly introduced from Europe, where this form is common"
(Man.) July.

R. aquatilis, L., var. trichophyllus, Gray. WHITE WATER-CROWFOOT.
Ashby, Wilmington, Malden, et al. Not uncommon. June-July.

R. multifidus, Pursh. YELLOW WATER-CROWFOOT.
Groton, Medford, Concord, et al. Rather common. May-June.

R. Flammula, L., var. reptans, Meyer. CREEPING SPEARWORT.
Lowell, Groton and Tewksbury (Dr. C. W. Swan); Concord and
Lexington (F. S. Collins); Reading (W. H. Manning). Widely
distributed, but not common. June-Aug.

R. Cymbalaria, Pursh. SEA-SIDE CROWFOOT.
Marshes; Cambridge (Bigelow's Fl. Bost.); Medford (Mrs. P. D.
Richards); Malden (F. S. Collins). Not very common. June-Aug.

R. abortivus, L. SMALL-FLOWERED CROWFOOT.
Common. May-June.

R. abortivus, L., var. micranthus, Gray.
Melrose (C. J. Sprague; Rev. Thos. Morong). United with the
type by a graded series of specimens. May-June.

R. sceleratus, L. CURSED CROWFOOT.
Somerville (C. E. Perkins); Cambridge (F. S. Collins); Belmont
(H. S. Richardson); Waltham List. Scarce. May-July.

R. recurvatus, Poir. HOOKED CROWFOOT.
Malden, Belmont, Lowell, et al. Scarce. May-June.

*R. Pennsylvanicus, L. f. BRISTLY CROWFOOT.
Concord, rare (Minot Pratt).

R. fascicularis, Muhl. EARLY CROWFOOT.
Common. April-May.

R. repens, L. CREEPING CROWFOOT.
Medford, Townsend, Concord, et al. Not uncommon. May-July.

R. BULBOSUS, L. BUTTERCUPS. BULBOUS CROWFOOT.
Very common. Forms with double flowers occasional. May-July. Nat. from Eu.

R. ACRIS, L. TALL BUTTERCUPS.
Very common. May-Aug. Nat. from Eu.

ISOPYRUM, L.

*I. biternatum. Torr. and Gray.
Concord; introduced from Michigan by Minot Pratt. May.

CALTHA, L.

C. palustris, L. MARSH MARIGOLD.
Widely known by the name of COWSLIPS, a totally different plant.
Frequent. April-May.

COPTIS, Salisb.

C. trifolia, Salisb. GOLDTHREAD.
Widely distributed, but not abundant. May.

AQUILEGIA, Tourn.

A. Canadensis, L. WILD COLUMBINE.
Common. May-June.

A. vulgaris, L.
The common GARDEN COLUMBINE of Europe. Concord, escaped, (Minot Pratt); et al. July.

DELPHINIUM, Tourn.

D. Consolida, L. FIELD LARKSPUR.
Stoneham, rubbish heap in woods, apparently spreading, August, 1885, (F. S. Collins). July-Aug. Int. from Eu.

XANTHORRHIZA, Marsh.

**X. APIIFOLIA, L'Her. YELLOW-ROOT.
Concord. Found growing by the roadside by Minot Pratt; locally established, but can hardly be native. Nat. from the South.

ACTÆA, L.

A. spicata, L. var. **rubra,** Ait. RED BANEBERRY.
Widely distributed, but nowhere common. May-June.

A. alba, Bigel. WHITE BANEBERRY. COHOSH.
Distribution as in the preceding. May-June.

BERBERIDACEÆ. BARBERRY FAMILY.

BERBERIS, L.

B. VULGARIS, L. BARBERRY.
Common. More abundant in the eastern section of the county. May-June. Nat. from Eu.

PODOPHYLLUM, L.

P. PELTATUM, L. MAY APPLE. MANDRAKE.
Shirley, (F. L. Sargent); Framingham, (Rev. J. H. Temple); Burlington, (Miss M. E. Carter). May-June. Nat. from farther west.

NYMPHÆACEÆ. WATER-LILY FAMILY.

BRASENIA, Schreb.

B. peltata, Pursh. WATER-SHIELD.
Common. July-Aug.

NELUMBIUM, Juss.

N. LUTEUM, Willd. YELLOW NELUMBO. WATER CHINQUEPIN.
Concord River, Concord, (Walter Deane; specimen in herb. of).
July-Aug. Introduced from the South.

NYMPHÆA, Tourn.

N. odorata, Ait. WHITE POND-LILY.
Common. June-Aug.
*N. odorata, Ait. var. **minor,** Sims.
Concord (H. S. Richardson).

NUPHAR, Smith.

N. advena, Ait. YELLOW POND-LILY. COW-LILY.
Common. May-Aug.
N. Kalmianum, Ait. (N. luteum, Smith, var. pumilum; Man.)
SMALLER COW-LILY.
Walden Pond, Concord (J. L. Russell, Hovey's Mag. Vol. XXI);
Sudbury (Bigelow's Fl. Bost.); Concord River, Tewksbury (C. W. Jenks). Not common. Aug.-Sept.

SARRACENIACEÆ, PITCHER-PLANT FAMILY.

SARRACENIA, Tourn.

S. purpurea, L. PITCHER-PLANT. SIDE-SADDLE FLOWER.
Rather common. May-June.

PAPAVERACEÆ, POPPY FAMILY.

ARGEMONE, L.

A. Mexicana, L. MEXICAN POPPY.
Cambridge; occasional in waste-heaps, with a white variety (Walter Deane; specimen in herb. of.) July-Sept. Int. from tropical America.

CHELIDONIUM, L.

C. MAJUS, L. CELANDINE.
Common near dwellings. Among the earliest plants introduced from Europe. May-July.

SANGUINARIA, Dill.

S. Canadensis, L. BLOOD-ROOT.
Generally distributed, but found in small patches. April-May.

FUMARIACEÆ. FUMITORY FAMILY.

ADLUMIA, Raf.

A. CIRRHOSA, Raf. FUMITORY.
Persistent for years in an old garden at Lowell (Dr. C. W. Swan); Concord; introduced from the West by Minot Pratt; now locally established.

DICENTRA, Bork.

*D. CUCULLARIA, DC. DUTCHMAN'S BREECHES.
Concord; introduced from the West by Minot Pratt; now locally established. Doubtfully reported elsewhere in the county. May.

CORYDALIS, Vent.

C. glauca, Pursh. PALE CORYDALIS.
Not uncommon. May-Aug.

*C. aurea, Wild. GOLDEN CORYDALIS.
"Near Nobscot, Framingham; found about fifteen years ago; not since reported." (Rev. J. H. Temple.)

FUMARIA, L.

F. OFFICINALIS, L. FUMITORY.
Lowell (Dr. C. W. Swan); Medford (L. L. Dame); Ashland (Rev. Thos. Morong); Natick (Austin Bacon). Persistent in old gardens. June-Aug. Nat. from Eu.

CRUCIFERÆ. MUSTARD FAMILY.

NASTURTIUM, R. Br.

N. OFFICINALE, R. Br. TRUE WATER-CRESS.
Somerville (C. E. Perkins); Groton (C. W. Jenks); Medford, abundant (L. L. Dame); Belmont, abundant (Walter Deane). June-July. Nat. from Eu.

N. sylvestre, R. Br. Yellow Cress.
Medford, 1887 (F. S. Collins); Newton (C. J. Sprague; Man. 1859).
May-June. Nat. from Eu.

N. palustre, DC. Marsh Cress.
Somerville, Cambridge, Framingham, Groton, et al. Rather common. May-July.

N. Armoracia, Fries. Horseradish.
Often found escaped from cultivation. Apparently persistent. May-June. Nat. from Eu.

N. amphibium, R. Br. var. *auriculatum*, Reich.
Cambridge, rubbish-heap, 1884 (Walter Deane; specimen in herb. of). Adv. from Eu.
"Pod elliptical or oblong, three or four times shorter than the pedicel; immersed leaves undivided, lanceolate, attenuate at both ends, sessile; upper leaves pectinato-pinnatifid or lyrate; petals longer than the calyx; *var. auriculatum*, DC. Leaves furnished with small auricles at the base." Koch, Syn. Flor. Germ.

Dentaria, L. Pepper-root.

D. diphylla, Michx.
Arlington (E. Tuckerman); Belmont, specimen in the Boott herb. May.

D. heterophylla, Nutt.
Lowell (Albert S. Guild; W. P. Atwood). Rare. Apr.-May. Adv. from the South.

* **D. laciniata**, Muhl.
Belmont, near Railroad Station (Wm. Boott). Apr.-May.

Cardamine, L.

C. rhomboidea, DC. White Spring-Cress.
Medford and Belmont (F. S. Collins); Woburn (C. E. Perkins); Framingham (Rev. J. H. Temple); et al. May.

* **C. rhomboidea**, DC., var. **purpurea**, Torr. Purple Spring-Cress.
Arlington, May 8, 1865 (Wm. Boott).

C. hirsuta, L. Small Bitter-Cress.
Common. May-June.

C. hirsuta, L., var. **sylvatica**, Gray.
Melrose and Medford (F. S. Collins). May-June.

Arabis, L.

A. hirsuta, Scop. Hairy Rock-Cress.
Somerville (C. E. Perkins); June-July.

A. laevigata, Poir. SMOOTH ROCK-CRESS.
Melrose and Medford (F. S. Collins); Pine Hill, Medford, 1885 (Dr. C. W. Swan); Woburn (Wm. Boott). Not common. May-June.

A. Canadensis, L. SICKLE-POD.
Medford, Chelmsford, Acton, et al. Not common.

A. perfoliata, Lam. TOWER MUSTARD.
Malden and West Medford (Wm. Boott); Billerica (Dr. C. W. Swan). Rare. June-July.

A. confinis, Wats. (A. Drummondii, Gray; Man.)
Dracut (Dr. C. W. Swan); Concord (Walter Deane). Rare. June-July.

BARBAREA, R. Br.

B. VULGARIS, R. Br. YELLOW ROCKET. WINTER CRESS.
Common. The forms known as var. STRICTA, Regel, Somerville and Tewksbury (Dr. C. W. Swan); Malden (F. S. Collins); and var. ARCUATA, Koch, Lowell (Dr. C. W. Swan). May-June. Thoroughly established, but probably naturalized from Eu.

ERYSIMUM, L.

E. cheiranthoides, L. WORM-SEED MUSTARD.
Medford, 1866 (Wm. Boott); Malden (F. S. Collins); along railroad, Lowell; near woollen mills, Westford (Dr. C. W. Swan); occasional along the B. & A. R. R., Ashland (Rev. Thos. Morong). June-July.

E. repandum, L.
Westford, woollen-mill yard, a few plants, 1884 and 1885 (Dr. C. W. Swan). July. Adv. from Eu.
"Stem 5-10 in. high. Leaves lanceolate, acuminate, crenate or repand-dentate, or entire, recurved at the tip, somewhat rough-hairy. Flowers yellow, pedicels half the length of the calyx. Pods obtusely four-angled, nearly terete, scarcely larger than the horizontally patent pedicel." Wagner, Deutsche Flora.

SISYMBRIUM, L.

S. OFFICINALE, Scop. HEDGE MUSTARD.
Common. June-Sept. Nat. from Eu.

S. Sophia, L.
Chelmsford, abundant on a small patch of waste ground in 1884, wholly gone in 1885 (Dr. C. W. Swan). June. Adv. from Eu.

S. Pannonicum, Jacq.
Westford, woollen-mill yard; Tewksbury, roadside, a few plants in a limited area (Dr. C. W. Swan). June-July,

"Lower leaves runcinate-pinnatifid, laciniæ dentate with auriculate base, auricle ascending; upper leaves pinnate, pinnæ all narrowly linear; sepals widely spreading; pods and pedicels of about equal diameter, spreading."
Koch, Syn. Flor. Germ.

S. Lœselii, L.
Westford, woollen-mill yard, 1884 and 1885 (Dr. C. W. Swan). July-Aug. Adv. from Eu.

"Stem 1-3 ft. high, hispid with stiff hairs, as are also the lower leaves. Leaves runcinate-pinnatifid, the terminal segment very large, hastate; calyx patent; pods ascending, twice the length of the spreading pedicels, the younger shorter than the convex cluster."
Wagner, Deutsche Flora.

S. incisum, Engelm.
N. Chelmsford, in wool-waste (Rev. W. P. Alcott; specimen in herb. of). Adv. from Cal.

BRASSICA, Tourn.

B. SINAPISTRUM, Boiss. CHARLOCK. YELLOW MUSTARD.
Westford, Medford, Ashland, et al. Rather common in waste grounds. June-Aug. Nat. from Eu.

B. ALBA, Gray. WHITE MUSTARD.
Groton and Dracut, waste-heaps (Dr. C. W. Swan); Somerville (C. E. Perkins). June-July. Nat. from Eu.

B. NIGRA, Koch. BLACK MUSTARD.
Persistent about dwellings as a relic of cultivation. Rather common. June-Aug. Nat. from Eu.

B. campestris, L. KALE.
Persistent in old gardens, and common on "dumps." June-July. Int. from Eu. The cultivated forms of *B. campestris*, B. Rapa, L. Turnip, and B. Napus, L. Rape, are often persistent. For description, see Wood's Bot. & Fl.

DRABA, L.

*****D. arabisans, Michx.**
Concord; introduced from Vermont by Minot Pratt. May-June.

D. Caroliniana, Walt.
Woburn (C. E. Perkins). Precise locality now unknown. May.

D. verna, L. WHITLOW GRASS.
West Medford (Wm. Boott). This rare plant has appeared every spring for many years in the gravelly paths of a garden. No other locality is known. Apr.-May.

ALYSSUM, Tourn.

A. calycinum, L.
Medford (G. E. Davenport); Somerville (C. E. Perkins). May. Adv. from Eu.

CAMELINA, Crantz.

C. sativa, Crantz. FALSE FLAX.
Malden (F. S. Collins); Westford, woollen-mill yard (Dr. C. W. Swan); Medford (Wm. Boott). June-July. Adv. from Eu.

CAPSELLA, Vent.

C. BURSA-PASTORIS, Mœnch. SHEPHERD'S PURSE.
Everywhere. Apr.-Nov. Nat. from Eu.

THLASPI, Tourn.

T. arvense. L. MITHRIDATE MUSTARD.
Somerville (C. E. Perkins); Lowell; Chelmsford, two localities, one in a field which had been cultivated; numerous plants in another which had been dressed with wool-waste; no evidence of permanent residence (Dr. C. W. Swan). May. Adv. from Eu.

LEPIDIUM, L.

L. Virginicum, L. WILD PEPPERGRASS.
Common. June-Sept.

L. RUDERALE, L.
Lowell and Chelmsford (Dr. C. W. Swan); Somerville and Cambridge (C. E. Perkins); Weston (F. S. Collins). Rather common in the eastern section of the county. The rare form with petaliferous flowers has been found at Malden. May-June. Nat. from Eu.

L. CAMPESTRE, R. Br.
Newton (C. J. Sprague); Cambridge (T. W. Harris, Hovey's Mag. Vol. VI, 1840); Malden; Melrose, not uncommon, 1884 and 1885, (F. S. Collins); Lowell, on the road-bed of the B. & M. R. R. (Dr. C. W. Swan). June. Nat. from Eu.

SENEBIERA, DC.

S. didyma, Pers.
Cambridge, in walks (L. H. Bailey, Jr.) "An immigrant from farther south." Man. May-June.

ISATIS, L.

I. tinctoria, L. WOAD.
Newton (C. J. Sprague), *fide* specimen in herb. B. S. N. H. Adv. from Eu. For description, see Wood's Bot. & Fl.

CAKILE, Tourn.

C. Americana, Nutt. SEA-ROCKET.
Medford (C. E. Perkins); Malden (F. S. Collins). Along the Mystic. July-Aug.

RAPHANUS, L.

R. RAPHANISTRUM, L. WILD RADISH. JOINTED CHARLOCK.
A common weed in cultivated fields. May-July. Nat. from Eu.

HESPERIS, L.

H. MATRONALIS. L. ROCKET.
Reading, escaped (W. H. Manning); Malden, persistent for ten years in one locality and spreading (F. S. Collins); Stoneham (B. F. Gordon); Arlington (L. L. Dame). Appears to be sparingly established. June. Nat. from Eu. For description, see Wood's Bot. & Fl.

TROPIDOCARPUM, Hook.

"Pod linear, flattened laterally, often one-celled by the disappearance of the narrow partition; valve carinate, one-nerved. Seeds in two rows, minute, flattened, not winged; cotyledons incumbent. Style short. A low, slender, hirsute, branching annual, with pinnately divided leaves, and yellow solitary axillary flowers." Bot. Cal.

T. gracile. Hook.
Wool-waste, N. Chelmsford (Rev. W. P. Alcott; specimen in herb. of). Adv. from Cal.
"Stems weak, 2 feet high or less; leaves pinnatifid or rarely 2-pinnatifid, with narrow or linear segments; flowers in the axils of the upper bract-like leaves; petals 1½ to 3 lines long, nearly twice longer than the obtuse sepals; pods 6 to 20 lines long, more than a line broad, pointed at both ends, ascending on slender spreading pedicels 10 to 20 lines long." Bot. Cal.

VIOLACEÆ. VIOLET FAMILY.

V. rotundifolia, Michx. ROUND-LEAVED VIOLET.
Townsend (Miss H. E. Haynes); Framingham, rare (Rev. J. H. Temple); Ashby, rare (L. L. Dame); Concord, introduced from Vermont by Minot Pratt. April-May.

V. lanceolata, L. LANCE-LEAVED VIOLET.
Common. April-June.

V. primulifolia, L. PRIMROSE-LEAVED VIOLET.
Less common than the preceding. April-June.

V. blanda, Willd. SWEET WHITE VIOLET.
Common. April-June.

V. ODORATA, L. ENGLISH VIOLET.
Ashland, sparingly naturalized (Rev. Thos. Moroug; specimen in herb. of). Int. from Eu.

V. palmata, L. HAND-LEAVED VIOLET.
Groton (C. W. Jenks); Melrose (Rev. Thos. Morong); Stoneham (G. E. Davenport); Fresh Pond, Cambridge, specimen in Gray Herb. April-June.
"The late Prof. Tuckerman long ago (1839) collected at Concord specimens which would surely pass for V. pedatifida, if from the valley of the Mississippi." "V. pedatifida is indeed probably only a marked geographical variety of V. palmata." Gray's Rev. N. A. Violets.

V. palmata, L., var. **cucullata,** Gray.
The common county form; including var. **cordata,** Cambridge, (T. W. Harris, Hovey's Mag., Vol. VI., 1840; C. E. Perkins, 1882). April-June.

V. sagittata, Ait. ARROW-LEAVED VIOLET.
Very common. April-June.

V. pedata, L. BIRD-FOOT VIOLET. HORSE-SHOE VIOLET.
Common. Often light blue or white. May-June.

V. canina, L., var. **Muhlenbergii,** Gray (V. canina, L., var. sylvestris, Regel, Man.) DOG VIOLET.
Widely distributed, but not common. May-July.

* V. rostrata, Muhl.
Concord, introduced from the North by Minot Pratt. June-July.

* V. striata, Ait. PALE VIOLET.
Concord, introduced from the West by Minot Pratt. May-July.

V. Canadensis, L.
Concord, introduced from the North by Minot Pratt; doubtfully reported elsewhere. July.

V. pubescens, Ait. (including var. **eriocarpa,** Nutt., of Man.) YELLOW VIOLET.
Throughout the county, but nowhere abundant. May.

V. TRICOLOR, L. PANSY. HEART'S EASE.
Somerville (C. E. Perkins); Ashland, naturalized about old houses (Rev. Thos. Morong). April-May. Nat. from Eu.

* V. cornuta, L.
Concord, introduced from Europe by Minot Pratt.
"Root fibrous, diffuse; stems ascending; leaves cordate-ovate, crenate, ciliate; stipules obliquely cordate, inciso-dentate, ciliate; sepals subulate; the subulate spur longer than the calyx." DC., Prodr. I. 301.

CISTACEÆ. ROCK-ROSE FAMILY.

HELIANTHEMUM, Tourn.

H. Canadense, Michx. FROST-WEED.
Common. June-Aug.

LECHEA, L. PINWEED.

L. major, Michx.
Rather common. Aug-Sept.

L. thymifolia, Pursh., (including L. maritima, Legg.)
Chelmsford (Dr. C. W. Swan); Newton (C. J. Sprague); Bedford (Mrs. C. M. Fitch). Aug.-Sept. Too near L. minor.

L. tenuifolia, Michx.
Malden, Winchester, Westford, et al. Rather common. Aug-Sept.
"Leaves of radical shoots lanceolate, much longer than broad. Generally low and spreading; capsules large and conspicuous. The inner sepals even when broad, have only a midrib, and no side veins; all the other species have more or less distinctly 3 veins or ribs, rising from the base. The outer sepals about equal the inner in length." W. H. Leggett in Torrey Bulletin, VI. 251.

L. minor, L., (including L. intermedia, Legg.? and L. minor var. intermedia, Legg.)
Westford, Bedford, Cambridge, et al. Rather common. July-Sept.

L. racemulosa, Lam.
Ashland and Melrose (Rev. Thos. Morong).
"Leaves of radical shoots elliptical, ovate, or oblong, not more than two or three times as long as broad. Easily distinguished by its slender spreading pedicels, oblong flowers, and broadly lanceolate stem leaves. Outer sepals shorter." W. H. Leggett in Torrey Bulletin, VI. 251.

DROSERACEÆ. SUNDEW FAMILY.

DROSERA, L.

D. rotundifolia, L. ROUND-LEAVED SUNDEW.
Not uncommon. July-Aug.

D. intermedia, Drev. and Hayne, var. **Americana**, DC. (D. longifolia, L., Man.)
Common. July-Aug.

HYPERICACEÆ. ST. JOHN'S-WORT FAMILY.

HYPERICUM, L.

H. ellipticum, Hook.
Medford, Tewksbury, Chelmsford, et al. Not uncommon. June-Aug.

H. PERFORATUM, L. ST. JOHN'S-WORT.
Common. June-Aug. Nat. from Eu.

H. maculatum, Walt. (H. corymbosum, Muhl., Man.)
Not uncommon. July-Aug.

H. mutilum, L.
Common. July-Sept.

H. Canadense, L.
Very common. June-Sept.

H. nudicaule, Walt. (H. Sarothra, Michx, Man.)
Very common. June-Sept.

ELODEA, Juss., Pursh.

E. campanulata, Pursh. (Elodes Virginica, Nutt., Man.)
Common. July-Aug.

ELATINACEÆ. WATER-WORT FAMILY.

ELATINE, L.

E. Americana, Arn.
Mystic Pond (C. E. Perkins); Fresh Pond (Bigelow's Fl. Bost., under Crypta minima); Arlington (Rev. Thos. Morong).

CARYOPHYLLACEÆ. PINK FAMILY.

DIANTHUS, L.

D. ARMERIA, L. DEPTFORD PINK.
Malden, Medford, Reading, et al. Rather scarce. July-Aug. Nat. from Eu.

D. *deltoides*, L. Ashland, spontaneous in old gardens (Rev. Thos. Morong).
"A low plant with perennial rootstock, producing a tuft of procumbent leafy shoots; the annual flowering stems erect or ascending, smooth or slightly hairy, 5 or 10 in. high, dichotomously branched above. Leaves a third of an inch long, green, smooth, obtuse, the upper somewhat acute. Flowers rather small, scentless, purple or whitish, spotted, singly or in pairs, on short pedicels. Sepals with

pointed teeth; the involucral bracts abruptly acuminate to an acute point, which reaches about one-third the length of the calyx.' Wagner, Deutsche Flora.

D. barbatus, L. SWEET WILLIAM.
Wilmington (F. S. Collins). Growing in a thicket, remote from dwellings. June to July. Adv. from Eu. For description, see Wood's Bot. & Fl.

SAPONARIA, L.

S. OFFICINALIS, L. SOAPWORT. BOUNCING BET.
Rather common. July-Sept. Nat. from Eu.

S. Vaccaria, L. (Vaccaria vulgaris, Host., Man.) COW HERB.
Lowell, "dumps" (Dr. C. W. Swan). Adv. from Eu.

GYPSOPHILA, L.

"Annual or perennial branching plants with entire, narrow leaves and numerous, white or reddish flowers, Calyx bell-shaped, 5-toothed or-cleft, without involucre at the base, with as many or three times as many longitudinal nerves as teeth. Petals wedge-shaped, narrowed in the claw, without scale. Capsule unilocular, opening with 4 teeth." Wagner, Deutsche Flora.

G. muralis, L.
Escaped from gardens, Bedford and Dracut (Dr. C. W. Swan).
"A delicate plant with stiffly erect stems 1-5 in. high, forking; flowers distant in an open panicle. Leaves linear, attenuate at both ends. Petals rose red with dark veins." Wagner, Deutsche Flora.

SILENE, L.

S. INFLATA, Smith. BLADDER CAMPION.
Common, at least in the eastern part of the county. June-July. Nat. from Eu.

S. Pennsylvanica, Michx. WILD PINK.
Not uncommon. May-June.

S. Armeria, L. SWEET WILLIAM CATCHFLY.
Reading (W. H. Manning); Medford (G. E. Davenport); Concord (Minot Pratt). Escaped from gardens. July. Adv. from Eu.

S. antirrhina, L. SLEEPY CATCHFLY.
Medford, Lowell, Billerica, et al. Not uncommon. May-July.

S. quinquevulnera, L.
Reading, escaped (W. H. Manning). Adv. from Eu. For description, see Wood's Bot. & Fl.

S. NOCTIFLORA, L. NIGHT-FLOWERING CATCHFLY.
Malden, Reading, Lowell. Ashland, et al. Widely distributed, but not very common. The Silene nocturna of Bigelow's Fl. Bost., and of Dewey's Report was probably this plant. June-Aug. Nat. from Eu.

S. apetala, Willd.
Lexington, growing with Anychia capillacea, on a wooded hillside, Aug. 6, 1883 (C. W. Jenks). Probably adv. from Eu.
"Hoary pubescent; stem erect, branching; leaves lanceolate, the upper linear; flowers few, terminal, or in the forks, calyx obovate, 10-striate; petals none." DC. Prodr., I. 369.

LYCHNIS, Tourn.

L. DIOICA, L.
Medford and Cambridge (C. E. Perkins); Ashland, established, (Rev. Thos. Morong). Scarce. Nat. from Eu. For description, see Wood's Class-Book.

L. vespertina, Sibth.
Cambridge, (*fide* specimen in Gray Herb.)

L. Githago, Lam. CORN COCKLE.
Westford (Miss Emily F. Fletcher); Lowell (Dr. C. W. Swan); Ashland, grain-field by railroad (Rev. Thos. Morong); Concord, cornfields (Minot Pratt). According to Dewey's Rep. on Herb. Plants, 1840, "scarcely naturalized, but propagated with the wheat;" no more evidence of naturalization at present. June-July.

ARENARIA, L.

A. SERPYLLIFOLIA, L. THYME-LEAVED SANDWORT.
Rather common. May-Sept. Nat. from Eu.

A. lateriflora, L.
Common eastward. May-June.

STELLARIA, L.

S. MEDIA, Smith. CHICKWEED.
Everywhere. April-Nov. Nat. from Eu.

S. longifolia, Muhl. LONG-LEAVED STITCHWORT
Lowell (Miss M. Swan). Scarce. June-July.

S. GRAMINEA, L.
Malden and Belmont (F. S. Collins); Watertown (L. H. Bailey, Jr.); Winchester (C. E. Perkins); Cambridge (Walter Deane); Ashland (Rev. Thos. Morong). Not uncommon. June-July. Nat. from Eu.
Often confounded with the preceding, but its affinities are rather with S. longipes of the Manual from which it differs in the always linear-lanceolate leaves (broadest above the base), the divaricate pedicels, and more elongated inflorescence.

S. uliginosa, Murr. SWAMP STITCHWORT.
Lowell, July 11, 1883 (Dr. C. W. Swan). Rare.

S. borealis, Bigel. NORTHERN STITCHWORT,
Ashby (Dr. C. W. Swan); Tewksbury (J. R. Churchill, 1884); Waltham (F. S. Collins). May-June.

S. aquatica, Scop.
"Newtonville, Aug., 1881, beside the R. R. track," (C. J. Sprague.) Specimen in herb. B. S. N. H. This should not be confounded with S. aquatica, Pollich, of the Man. 1st Ed., which is S. uliginosa, Murr. Adv. from Eu.
" Stem diffusely branched. decumbent. Flowering stems branched below the cyme. Leaves ovate, acute or acuminate; the lower ones on footstalks shorter than the laminae, the middle and upper ones sessile. Flowers numerous, in dichotomous cymes terminating the stem and branches. Sepals lanceolate, rather obtuse, faintly 1-nerved, with broad scarious margins, the herbaceous part with short gland-tipped hairs. Fruit stalks spreading or reflexed. Capsule drooping, longer than the sepals, ovate-conical. Stem with short gland-tipped hairs." Sowerby, Eng. Bot.. II. 91,

CERASTIUM, L.

C. viscosum, L. MOUSE-EAR CHICKWEED.
Common. April-Sept. Nat. from Eu.

C. arvense, L. FIELD CHICKWEED.
Concord (H. S. Richardson); Ashland (Rev. Thos. Morong); Framingham (Miss J. W. Williams); Medford (C. E. Perkins). Rare. May-June.

SAGINA, L.

S. procumbens, L. PEARLWORT.
Waltham (C. E. Perkins); Ashland (Rev. Thos. Morong); Medford (F. S. Collins). Sometimes growing in brick sidewalks; scarce. June-July.

LEPIGONUM, Fries.

L. rubrum, Fries. (Spergularia rubra, Presl, var. campestris, Man.)
Rather common. May-Aug.

L. salinum, Fries, (Spergularia salina, Presl, Man.)
Salt marshes; Cambridge (C. E. Perkins); Medford (F. S. Collins). July.

L. medium, Fries. (Spergularia media, Presl, Man.)
Salt marshes; Medford and Everett (F. S. Collins). July.

SPERGULA, L.

S. ARVENSIS, L. CORN SPURREY.
Lowell, Cambridge, Concord, et al. Not uncommon. June-Aug. Nat. from Eu.

PARONYCHIÆ. WHITLOW-WORT FAMILY.

ANYCHIA, Michx.

A. capillacea, DC. (A. dichotoma of Man., in part.)
Lexington (Dr. C. W. Swan); et al. July.

SCLERANTHUS, L.

S. annuus, L. Knawel.
Common. June-July.

FICOIDEÆ.

MOLLUGO, L.

M. VERTICILLATA, L. CARPET-WEED.
A very common weed. June-Sept. Nat. from the South.

PORTULACACEÆ. PURSLANE FAMILY.

PORTULACA, Tourn.

P. OLERACEA, L. PURSLANE; the " Pusley" of the farmer.
In cultivated land everywhere. Still used to some extent as "table greens." July-Sept. Nat. from Eu.

P. pilosa, L.
Lowell, escaped (Dr. C. W. Swan). Int. from the South.
"Leaves linear, obtuse, with a tuft of hairs in the axils; flowers purple; stamens about 20." Chapman's S. Fl.

CLAYTONIA, L. SPRING BEAUTY.

*****C. *Virginica,* L.
Concord, introduced from Indiana by Minot Pratt. April-May.

C. Caroliniana, Michx.
Ashby (E. Adams Hartwell). The Concord plant was introduced from Vermont by Minot Pratt. April-May.

MALVACEÆ. MALLOW FAMILY.

MALVA, L.

M. ROTUNDIFOLIA, L. MALLOW.
Common. June-Sept. Nat. from Eu.

M. crispa, L. CURLED MALLOW.
Reading (R. Frohock); Ashland, spontaneous (Rev. Thos. Morong). Adv. from Eu.

M. moschata, L. MUSK MALLOW.
Lowell, roadside (Dr. C. W. Swan); Malden (R. Frohock). July-Sept. Adv. from Eu.

M. Alcea, L.
Dracut, roadside; Lowell, railroad bed; Hopkinton, roadside (Dr. C. W. Swan); Medford (Mrs. P. D. Richards). Tending to establish itself. Adv. from Eu.

M. BOREALIS, Wallm.
Lowell, Dracut and Westford, near woollen mills, (Dr. C. W. Swan); E. Cambridge, Sept. 12, 1881 (C. E. Perkins); Cambridge, 1884 (Walter Deane). A native of Eu., but introduced in California wool, and so common in the vicinity of woollen mills that it may fairly claim naturalization. Aug.-Sept.
"Annual, erect or somewhat decumbent, hairy or nearly glabrous; leaves round-cordate, crenate, more or less strongly 5-7 lobed; peduncles axillary, solitary or clustered, 1 to 3 lines long; calyx-lobes acute, becoming very broad and enlarged in fruit; petals 2 or 3 lines long; carpels transversely reticulate-rugose." Bot. Cal.

SIDA. L.

S. spinosa, L.
Watertown (C. E. Perkins); Lowell, "dump" (Dr. C. W. Swan); Malden (F. S. Collins). Aug.-Sept. Probably introduced in Southern cotton.

ABUTILON, Tourn.

A. Avicennæ, Gærtn. VELVET-LEAF.
Somerville, Cambridge, Bedford, et al. Tending to establish itself. Aug.-Sept. Adv. from India.

HIBISCUS, L.

H. moscheutos, L. SWAMP ROSE-MALLOW.
Widely distributed, but not abundant. Aug.-Sept.

H. Trionum, L. BLADDER KETMIA.
Ashland, occasionally escaped from gardens (Rev. Thos. Morong). Adv. from Eu.

TILIACEÆ. LINDEN FAMILY.

TILIA, L.

T. Americana, L. BASSWOOD. WHITE-WOOD. LINDEN.
Rather common. June.

LINACEÆ. FLAX FAMILY.

LINUM, L.

L. Virginianum, L.
Malden, Melrose, Framingham, et al. Not very common. July-Sept.

L. sulcatum, Ridd.
Arlington (Wm. Boott). Specimen in the Boott Herb. Very rare.

L. usitatissimum, L. COMMON FLAX.
Occasional on "dumps" and along the roadside. June-July. Origin unknown; a weed of cultivation the world over. For description, see Wood's Bot. & Fl.

GERANIACEÆ. GERANIUM FAMILY.

GERANIUM, L.

G. maculatum, L. CRANESBILL.
Common. May-July.

G. Carolinianum, L. CAROLINA CRANESBILL.
Malden, Medford, Groton, et al. Not very common. A loose-flowering form with long peduncles and pedicels is found in Middlesex Fells, probably the plant credited to the same locality in Bigelow's Fl. Bost., as G. dissectum. June-Aug.

G. dissectum, L. CUT-LEAVED GERANIUM.
Lowell, "dumps" (Dr. C. W. Swan). Adv. from Eu.

G. Robertianum, L. HERB ROBERT.
Malden, Melrose, Groton, et al. Not uncommon eastward. June-Oct.

ERODIUM, L'HER.

E. CICUTARIUM, L'Her.
Chelmsford and Dracut, woollen mill yards (Dr. C. W. Swan); N. Chelmsford, abundant and spreading, 1878 and 1880 (Rev. W. P. Alcott); Winchester, 1885-6 (Mrs. P. D. Richards). Persistent, at least for several years; the common Erodium of the wool-waste; seems to have made a permanent settlement. Aug. Nat. from Eu.

E. Botrys, Bertol.
Westford, woollen-mill yard (Dr. C. W. Swan). A native of South Europe, but introduced in California wool.
"Leaves oblong, pinnatifid; the lobes dentate, obtuse; stipules small; sepals 4 lines long; beaks of the carpels 2 or 3 inches long." Bot. Cal.

E. moschatum, Willd.
Lowell and Westford, near woollen mills; Chelmsford, numerous thrifty plants in a garden dressed with wool-waste (Dr. C. W. Swan). Adv. from Eu. via California.
"Leaves pinnate, the oblong ovate leaflets unequally and doubly serrate; stipules conspicuous; pedicels mostly shorter and stouter, (than in E. cicutarium); sepals larger, 3 or 4 lines long; odor musky Bot. Cal.

IMPATIENS, L.

I. fulva, Nutt. JEWEL-WEED.
Common. June-Sept.

OXALIS, L.

O. Acetosella, L. WOOD-SORREL.
Ashby, (E. Adams Hartwell); Concord, 1862 (Horace Mann). As this plant was introduced into Concord by Minot Pratt, it is probable that Mann's specimen came from this source. June. Very rare.

O. violacea, L. VIOLET WOOD-SORREL.
Newton (E. B. Kenrick, Hovey's Mag., 1836); Concord, 1862 (Horace Mann); Belmont, rare, open field, 1883 (Walter Deane); Weston, rather common, 1887 (Walter E. Coburn). May-June.

O. corniculata, L., var. **stricta**, Sav. (O. stricta, L., Man.)
Common. May-Sept.

RUTACEÆ. RUE FAMILY.

XANTHOXYLUM, Colden.

X. Americanum, Mill. NORTHERN PRICKLY ASH.
Medford, Chelmsford, Westford, Acton, et al. Possibly introduced. April-May.

PTELEA, L.

P. trifoliata, L. HOP-TREE.
Somerville, Medford, et al. Occasionally spontaneous. From farther South. June.

SIMARUBACEÆ.

AILANTHUS, Willd.

A. GLANDULOSUS, Desf. TREE OF HEAVEN.
In many places spreading from seed and by suckers. June. Nat. from China.

ANACARDIACEÆ. CASHEW FAMILY.

Rhus, L. Sumach.

R. typhina, L. Staghorn Sumach.
Common. June-July.

R. glabra, L. Smooth Sumach.
Common. The form var. **laciniata,** Carriere, has been found along the F. R. R. in Weston. June-July.

R. copallina, L. Dwarf Sumach.
Rather common. July.

R. venenata, DC. Poison Sumach. Dogwood.
Common. June.

R. Toxicodendron, L. Poison Ivy. Mercury.
Both the upright and running forms are very common. June.

R. Cotinus, L.
Often found escaped, but can scarcely be considered naturalized. June. Int. from Eu.

VITACEÆ. VINE FAMILY.

Vitis, Tourn.

V. Labrusca, L. Northern Fox Grape.
Common. A variety with white fruit occasional. June.

V. æstivalis, Michx. Summer Grape.
Groton (C. W. Jenks); Medford (George E. Davenport); Concord, rare (Minot Pratt); Weston (L. L. Dame). Less common than V. Labrusca. June.

V. riparia, Michx. (V. cordifolia, Michx., var. riparia, Man.)
Sudbury (Dr. C. W. Swan). The V. cordifolia reported at Concord by Minot Pratt is probably this species. Rare. May.

Ampelopsis, Michx.

A. quinquefolia, Michx. Woodbine. Virginia Creeper.
Common. July.

Rhamnaceæ. Buckthorn Family.

Rhamnus, Tourn.

R. cathartica, L. Buckthorn.
Medford, Framingham, Ashland, et al. May-June. Nat. from Eu.

Ceanothus, L.

C. Americanus, L. New Jersey Tea.
Common. July.

CELASTRACEÆ. STAFF-TREE FAMILY.

CELASTRUS, L.

C. scandens, L. ROXBURY WAXWORK. CLIMBING BITTER-SWEET.
Rather common. June.

SAPINDACEÆ. SOAPBERRY FAMILY.

STAPHYLEA, L.

S. trifolia, L. BLADDER-NUT.
Weston (Bigelow's Fl. Bost.); not since reported in Middlesex, but found (1886) in the adjacent town of Needham, in Norfolk Co. May-June.

ÆSCULUS, L.

Æ. HIPPOCASTANUM, L. HORSE-CHESTNUT.
Propagates itself occasionally by seed. May-June. Int. from Asia via Eu.

CARDIOSPERMUM, L.

C. Halicacabum, L.
Somerville (C. E. Perkins). Adv. from the Southwest.
For description, see Wood's Bot. & Fl.

ACER, Tourn.

A. Pennsylvanicum, L. STRIPED MAPLE.
Concord (Minot Pratt); Ashby (L. L. Dame). Rare. May-June.

A. spicatum, Lam. MOUNTAIN MAPLE.
Concord, int. from Wachusett by Minot Pratt; Ashby (L. L. Dame). The elevated region about Mt. Watatic seems to be the only county station where this maple grows naturally. June.

A. saccharinum, Wang. ROCK MAPLE. SUGAR MAPLE.
Not abundant. Apr.-May.

A. dasycarpum, Ehrh. WHITE MAPLE.
Lowell, Tewksbury, Bedford, et al. Not common native. March-Apr.

A. rubrum, L. RED MAPLE. SWAMP MAPLE.
Very common. Apr.

POLYGALACEÆ. MILKWORT FAMILY,

POLYGALA, Tourn.

P. sanguinea, L.
Common. July-Aug.

P. cruciata, L.
Reading, Woburn, Bedford, et al. Not abundant. July-Sept.
P. verticillata, L.
Not uncommon. July-Aug.
P. polygama, Walt.
Malden (F. S. Collins); Medford (G. E. Davenport); et al. Not common except in the eastern towns. July-Aug.
P. paucifolia, Willd. FRINGED POLYGALA.
Widely distributed but not abundant. A form with white flowers, discovered by Henry M. Pratt at Concord, thrives and spreads at the expense of the type. May.

LEGUMINOSÆ. PULSE FAMILY.

LUPINUS, Tourn.

L. perennis, L. WILD LUPINE.
Rather common, except in the eastern towns. Flowers sometimes white. May-June.

CROTALARIA, L.

C. sagittalis, L. RATTLE-BOX.
Chelmsford (Dr. C. W. Swan); Cambridge (Bigelow's Fl. Bost.); Winchester, 1853 (Wm. Boott) station still existing; Concord (Minot Pratt); Weston, fields, (John L. Russell, 4th Mass. Rep. Agr.) Not common. July-Aug.

GENISTA, L.

G. TINCTORIA, L. WOAD-WAXEN. DYER'S WEED.
Malden, scarce, but gradually coming in from Essex Co. (F. S. Collins); Billerica (C. W. Jenks); Framingham (Miss J. W. Williams); Concord, 40 years ago a patch a rod in diameter, now covering half an acre (F. G. Pratt); Cambridge, a large patch has existed for years in an open field off Broadway (Walter Deane), probably the same locality mentioned by Tuckerman, 1841 (Notes, Josselyn Rareties). June-July. Nat. from Eu.

TRIFOLIUM, L.

T. ARVENSE, L. RABBIT-FOOT CLOVER.
Common. July-Aug. Nat. from Eu.
T. PRATENSE, L. RED CLOVER.
Everywhere. Sometimes found with white flowers. May-Sept. Nat. from Eu.

T. HYBRIDUM, L. ALSYKE.
This plant, occasional throughout the county, has within a few years become thoroughly established in the eastern and southern sections. June-Aug. Nat. from Eu.

"Heads roundish, dense; peduncles axillary, twice the length of the leaves; pedicels deflexed after flowering, the inner twice or thrice the length of the calyx tube; calyx smooth with naked throat, half the length of the corolla, with subulate teeth, the two upper longer; stems ascending, very smooth, hollow; stipules ovate, attenuate to a very acute point; leaves rhomboidal-elliptic, obtuse, serrulate. Lower leaves obovate; flowers white to rose-color." Koch, Syn. Fl. Germ.

T. REPENS, L. WHITE CLOVER.
Very common. Possibly indigenous, but probably introduced from Eu. May-Sept.

T. AGRARIUM, L. YELLOW OR HOP-CLOVER.
Widely distributed, but not very common. June-Aug. Nat. from Eu.

T. PROCUMBENS, L. LOW HOP-CLOVER.
Medford, Groton, et al. Not so common as the preceding. July. Nat. from Eu. Form known as var. MINUS, Koch, occasional.

T. *Dalmaticum*, Vis.
Lowell, "dump" (Dr. C. W. Swan). Adv. from Eu.

"Heads terminal and axillary, the axillary sessile or nearly so; calyx naked at the base; stipules dilated, at least the upper; calyx teeth not longer than the tube; calyx sulcate in fruiting; flowers red; stems decumbent, with appressed hairs." Cesati, Passerini and Gibelli, Flora Italiana.

T. *Macraei*, Hook and Arn.
Wool-waste, N. Chelmsford (Rev. W. P. Alcott). Adv. from Cal.
"Somewhat villous, with appressed or spreading hairs, erect, slender, a half to a foot high; stipules ovate to lanceolate; leaflets obovate to narrowly oblong, obtuse or retuse, serrulate, about half an inch long; flowers dark purple, 3 lines long, in dense ovate long-peduncled heads, calyx very villous; the straight teeth as long as the petals, often tinged with purple; pod 1-seeded." Bot. Cal.

MELILOTUS, Tourn.

M. *parviflora*, Desf.
Lowell, "dumps;" Westford, woollen-mill yards (Dr. C. W. Swan).

"Annual, smooth, erect, often 2 or 3 feet high, branching; leaflets mostly cuneate-oblong, obtuse, denticulate, an inch long or less;

flowers yellow, a line long, nearly sessile. Native of the Mediterranean region, now widely naturalized in warm countries, and common in California. Cattle are fond of it." Bot. Cal.

M. OFFICINALIS, Willd. YELLOW MELILOT.
Groton, Cambridge, Malden, et al. Occasional in waste grounds. June-Sept. Nat. from Eu.

M. ALBA, Lam. SWEET CLOVER. WHITE MELILOT.
Malden, Medford, Westford, et al. Sometimes found growing in dense patches in out of the way places, apparently well established, June-Sept. Nat. from Eu.

MEDICAGO, L.

M. sativa, L. LUCERNE. ALFALFA.
Cambridge, spontaneous in vegetable gardens; Malden, spreading beyond garden limits (Sylvester Baxter). This plant has a tendency to die out when cultivation ceases, and can scarcely be said to have effected a settlement. July-Aug. Int. from Eu.

M. LUPULINA, L. BLACK MEDICK. NONESUCH.
Common in the eastern part of the county. June-Sept. Nat. from Eu.

M. maculata, Willd. SPOTTED MEDICK.
Lowell, woollen-mill yards; Chelmsford, in a field dressed with wool-waste (Dr. C. W. Swan); Somerville (C. E. Perkins); N. Chelmsford, wool-waste (Rev. W. P. Alcott). This was erroneously given as M. denticulata in Mr. Alcott's list. Adv. from Eu.

M. denticulata, Willd.
Lowell, Billerica, and Dracut, woollen-mill yards; the commonest Medick of the mill yard, except possibly M. lupulina (Dr. C. W. Swan). Adv. from Eu.

M. lappacea, Lam.
Westford, wool-waste (Dr. C. W. Swan). Adv. from Eu.
"Smoothish; stems procumbent; leaflets obcordate, dentate; stipules ciliate-dentate; peduncles 3-5 flowered; pods spiral, smooth, of three whorls, obliquely flexuous-nerved, margins aculeate, spines long, hooked; seeds reniform, subtruncate, yellow." DC. Prodr. II. 177.

M. laciniata, All.
Lowell, "dumps" (Dr. C. W. Swan). Adv. from Eu.
"Stem erect; leaflets linear, incised-dentate, truncate; stipules ciliate-dentate; peduncles 1-2-flowered; pods spiral, sub-globose, very spiny, spines erect, subulate, hooked, compressed-canaliculate; margin thick, not sulcate nor zoned; seeds oblong-reniform." DC. Prodr. II. 180.

M. minima L.

Lowell, " dumps;" Westford, woollen-mill yard (Dr. C. W. Swan);
N. Chelmsford, wool-waste (Rev. W. P. Alcott). Adv. from Eu.

"Peduncles 1-2-flowered, longer than the leaves, shorter than the spiral, sub-globose, spiny, slightly hairy pods; whorls about 5, loose, veinless, with narrow, obtuse, distichously spiny margin; spines patent, subulate, straight with a hooked point, furrowed on each side; stipules ovate, shortly denticulate, the upper subentire; leaves obovate, denticulate; petioles, peduncles and stem pubescent." Koch, Syn. Fl. Germ.

M. præcox, DC.

Westford, wool-waste (Dr. C. W. Swan). Adv. from Eu.

"Stems prostrate; leaves obcordate, denticulate; stipules ciliate-dentate; peduncles 1-2-flowered, short; pods spiral, smooth, plane on both sides; whorls slender, rugose, with thick, nerveless, plane margin, spiny at the sides; spines subdivergent, somewhat hooked at the apex; seeds ovate-reniform." DC. Prodr. II. 178.

M. aculeata, Willd.

Westford, wool-waste (Dr. C. W. Swan).

"Leaves rhombic-ovate, dentate; stipules dentate; peduncles about 2-flowered; pods spiral, cylindric, flat at the sides; the whorls muricate on the margin." "*Patria ignota.*" DC. Prodr. II. 179

M. intertexta, Willd.

Westford, wool-waste (Dr. C. W. Swan). Adv. from Eu. For description see Wood's Bot. and Fl.

TRIGONELLA, L.

"This genus is distinguished from MEDICAGO by the never spirally twisted pod. The species with small, ovate-oblong pods are distinguished from MELILOTUS by the beak." Boiss. Fl. Or.

T. Cassia, Boiss.

Lowell, waste-heap (Dr. C. W. Swan). Adv. from Asia Minor.

"Sparingly hairy; short prostrate stems; leaflets, cuneate-truncate or retuse, toothed at the apex; flowers solitary; lobes of calyx somewhat hairy, linear-lanceolate, $\frac{1}{2}$ as long as tube; corolla pale violet $1\frac{1}{2}$ times as long as calyx; pod smooth, cylindrical, scarcely curved, terminating in a beak about $\frac{1}{4}$ its own length, traversed longitudinal anastomsing nerves." Boiss. Fl. Or.

PETALOSTEMON, Michx.

P. violaceus, Michx. PRAIRIE CLOVER.

Concord, introduced from Mich. by Minot Pratt.

ROBINIA, L.

R. PSEUDACACIA, L. LOCUST.
Rather common. June. Nat. from the Middle States.

R. VISCOSA, Vent. CLAMMY LOCUST.
Lowell, escaped (Dr. C. W. Swan); Ashland, sparingly naturalized (Rev. Thos. Morong); Littleton, escaped (L. L. Dame). Slowly spreading in favorable localities. June. Int. from Va.

R. HISPIDA, L. BRISTLY LOCUST. ROSE ACACIA.
Ashland, sparingly naturalized (Rev. Thos. Morong); Malden, where it has formed dense thickets, and is spreading (F. S. Collins). June. From the South.

TEPHROSIA, Pers.

T. Virginiana, Pers. GOAT'S RUE. TEPHROSIA.
Not found in the eastern towns, and nowhere very abundant. July.

DESMODIUM, DC.

D. nudiflorum, DC.
Common. Scape occasionally leafy. Aug.

D. acuminatum, DC.
Not uncommon. July-Aug.

D. rotundifolium, DC.
Not uncommon in the eastern part of the county. Aug.

D. rotundifolium, DC., var. glabratum, Gray.
Woods in Waltham (Bigelow's Fl. Bost., under Hedysarum humifusum). County specimen in herb. Edwin Faxon. Aug.

*D. canescens, DC.
Roadside, Arlington, Aug., 1853 and 1869 (Wm. Boott).

D. cuspidatum, Hook.
Lowell (Dr. C. W. Swan); Woburn (L. L. Dame). Rare. July-Aug.

D. Dillenii, Darl.
Medford (Mrs. P. D. Richards); Westford (Misses Fletcher and Hodgman); Lowell, Dracut and Townsend (Dr. C. W. Swan); Malden and Wilmington (F. S. Collins). Aug.

D. paniculatum, DC.
Widely distributed, but not very common. July.

D. Canadense, DC.
Rather common. A form with white flowers reported at Newton by W. H. Manning. July-Aug.

D. rigidum, DC.
Not uncommon in the northern and western towns of the county; reported also at Woburn by Mrs. P. D. Richards. Aug.

D. ciliare, DC.
Bedford and Dracut (Dr. C. W. Swan); Concord (Minot Pratt). Rare. Aug.

D. Marilandicum, Boott.
Widely distributed, but not common. Aug.

LESPEDEZA, Michx. BUSH CLOVER.

L. repens, Bart.; including L. procumbens, Michx., Man.
Medford (G. E. Davenport); Malden (H. L. Moody); Framingham (Rev. J. H. Temple); Waltham (Wm. Boott). Aug.-Sept.

L. violacea, Pers. (L. violacea, Pers., var. divergens, Man.)
Malden (F. S. Collins); Woburn (L. L. Dame). Aug.

L. reticulata, Pers. (L. violacea, Pers., var. sessiliflora, Man.)
Rather common. Aug.

L. reticulata, Pers., var. **augustifolia,** Maxim. (L. violacea, Pers., var. augustifolia, Man.)
Lowell, Groton, Westford and Woburn (Dr. C. W. Swan); Medford (G. E. Davenport); Newton (Wm. Boott). Aug.

L. Stuvei, Nutt.
Hopkinton (Dr. C. W. Swan); Bedford (Edwin Faxon); Arlington (Mrs. P. D. Richards); Westford (Miss Emily F. Fletcher). Not common. Aug.

L. hirta, Ell.
Common. Aug.-Sept.

L. capitata, Michx.
Common. Aug.-Sept.

VICIA, Tourn.

V. SATIVA, L. COMMON VETCH. TARE.
Rather common. June-July. Nat. from Eu.

V. tetrasperma, Lois.
Lowell, "dumps" (Dr. C. W. Swan); Arlington (Mrs. P. D. Richards). Adv. from Eu.

V. hirsuta, Koch.
Lowell, "dumps" (Dr. C. W. Swan). Adv. from Eu.

V. Cracca, L.
Dracut, roadside (Dr. C. W. Swan); Everett and Malden (F. S. Collins); Concord (Thoreau); et al. Not very common. July.

LATHYRUS, L.

L. palustris, L. MARSH VETCHLING.
Cambridge (C. E. Perkins); Medford (F. S. Collins); Wilmington (Wm. Boott). June-July.

L. sativus, L. CHICK PEA.
Lowell "dump" (Dr. C. W. Swan). Adv. from Eu.
For description, see Wood's Bot. & Fl.

PISUM, L.

P. sativum, L.
Lowell "dump" (Dr. C. W. Swan). Adv. from Eu.
For description, see Wood's Bot. & Fl.

APIOS, Boerh.

A. tuberosa, Moench. GROUND-NUT. WILD BEAN.
Common. July-Aug.

AMPHICARPÆA, Ell.

A. monoica, Ell. HOG PEANUT.
Not uncommon. July-Aug.

BAPTISIA, Vent.

B. tinctoria, R. Br. WILD INDIGO.
Very common. June-Aug.

CASSIA, L.

C. Marilandica, L. WILD SENNA.
Widely distributed, but not common. July-Aug.
C. Chamæcrista, L. PARTRIDGE PEA.
Arlington, Everett, Medford, Littleton, et al.
Not common. Aug.
C. nictitans, L. WILD SENSITIVE PLANT.
Chelmsford, Winchester, Holliston, Ashland, et al.
Not common. July-Aug.

CORONILLA, L.

C. varia, L.
Westford, woollen-mill yard (Dr. C. W. Swan). June-July. Adv. from Eu. Likely to become established.
For description, see Wood's Bot. and Fl.

SCORPIURUS, L.

"Calyx short, campanulate, 5-toothed; the two upper teeth connate beyond the middle, sub-bilabiate; keel acuminate-rostrate; stamens diadelphous, the single filament dilated at the apex; pod elongate, circinately revolute, longitudinally furrowed, of 3-6 joints, one seeded." Koch, Syn. Fl. Germ.

S. subvillosa, L.
Westford, near woollen-mills (Dr. C. W. Swan). Adv. from Eu. "Pod smooth, the interior ribs entire, the exterior bearing 6-8 stiff spines, somewhat hooked at the apex." Koch, Syn. Fl. Germ.

ROSACEÆ. ROSE FAMILY.

PRUNUS, Tourn.

P. Americana, Marsh. WILD YELLOW or RED PLUM.
Concord (Minot Pratt); Medford (L. L. Dame); Cambridge (fide specimen in Gray Herb.); E. Lexington (C. W. Wellington). Occasional; fruit mostly dropping before maturity, or developing into monstrosities. May.

P. maritima, Wang. BEACH PLUM.
Chelmsford, Tyngsboro and Tewksbury (Dr. C. W. Swan); Wilmington (L. L. Dame). Rare. May. Possibly introduced from the sea-coast.

P. SPINOSA, L., var. INSITITIA, Gray. BULLACE PLUM.
Found " in the woods near Mt. Auburn," by Oakes, whose specimen is in the Gray Herb.; reported growing "on the banks of the Charles," in Emerson's Trees and Shrubs, 1846; and observed in 1885 by C. F. Batchelder, in the same locality, "scattered along at intervals for some distance." May. Nat. from Eu.

P. pumila, L. DWARF CHERRY.
Groton, Chelmsford, Medford, et al. Not very common. May-June.

P. Pennsylvanica, L. WILD RED CHERRY.
Not uncommon; abundant in the N. W. towns. May.

P. Virginiana, L. CHOKE-CHERRY.
Common. May-June.

P. serotina, Ehrh. WILD BLACK CHERRY.
Common. May-June.

P. AVIUM, L. ENGLISH CHERRY.
Often spontaneous in copses, and apparently established. May.

NEILLIA, Don.

N. opulifolia, Benth. & Hook. (Spiræa opulifolia, L., Man.) NINE-BARK.
Cambridge (C. E. Perkins); Melrose (L. L. Dame). Escapes. June.

SPIRÆA, L.

S. salicifolia, L. MEADOW-SWEET.
Very common. July-Aug.

S. tomentosa, L. HARDHACK.
Very common. A white variety in Westford (C. W. Jenks). July-Aug.

POTERIUM, L.

P. Canadense, Benth. & Hook. CANADIAN BURNET.
Dracut (Dr. C. W. Swan); Littleton (C. E. Perkins). Scarce.

AGRIMONIA, Tourn.

A. Eupatoria, L. AGRIMONY.
Not uncommon, July-Sept.

GEUM, L. AVENS.

G. album, Gmel.
Rather common. July-Aug.

G. Virginianum, L.
Malden (F. S. Collins); Cambridge (C. E. Perkins); Ashland (Rev. Thos. Morong). June-July.

G. strictum, Ait.
Somerville (Bigelow's Fl. Bost.); Chelmsford and Tewksbury (Dr. C. W. Swan); Medford (Mrs. P. D. Richards); et al. July.

G. rivale, L. WATER OR PURPLE AVENS.
Not uncommon. May-June.

G. triflorum, Pursh. "Called in the West, OLD MAID'S FRIZZLES." Concord, introduced from Wisconsin by Minot Pratt. May.

POTENTILLA, L. CINQUE-FOIL.

P. Norvegica, L.
Rather common. July-Sept.

P. Canadensis, L. COMMON CINQUE-FOIL. FIVE-FINGER.
Very common. Apr.-July.

P. Canadensis, L., var. **simplex,** Torr. & Gray.
Very common. May-July.

P. argentea, L. SILVERY CINQUE-FOIL.
Common. June-July.

P. arguta, Pursh.
Westford (Misses Fletcher and Hodgman); Chelmsford (Dr. C. W. Swan); Melrose (Rev. Thos. Morong); Concord (Thoreau). Not very common. June.

P. Anserina, L. SILVER-WEED.
Marshes in Malden, Medford, Everett and Cambridge. June-July.

P. fruticosa, L. SHRUBBY CINQUE-FOIL,
Reading, rare (W. H. Manning); Concord, rare (Minot Pratt); Groton, abundant (C. W. Jenks). June-Aug.

P. tridentata, Soland. THREE-TOOTHED CINQUE-FOIL.
Summit of Mt. Watatic, Ashby (L. L. Dame). The Concord plant was introduced from the White Mts. by Minot Pratt. June.

P. palustris, Scop. MARSH FIVE-FINGER.
Lowell (Dr. C. W. Swan); Malden and Stoneham (F. S. Collins); Cambridge (Geo. P. Huntington); So. Natick (Rev. Thos. Morong). Not very common. June-July.

P. recta, Willd.
Found for several seasons in Malden by F. S. Collins; now extinct in this station; has grown for a dozen or more years in Concord, near Monument street; scarcely established. Adv. from eastern Eu. and Asia. For description, see Wood's Class-Book.

FRAGARIA, Tourn. STRAWBERRY.

F. Virginiana, Duchesne.
Common. May.

F. vesca, L.
Common in the north-western, but infrequent in the eastern sections of the county. May-June.

DALIBARDA, L.

D. repens, L. (Rubus Dalibarda, L.)
Ashby, near base of Mt. Watatic (L. L. Dame). The Concord plant was introduced from Vermont by Minot Pratt. June-July.

RUBUS, Tourn.

R. odoratus, L. PURPLE FLOWERING RASPBERRY.
Lowell (Dr. C. W. Swan); Concord (F. S. Collins); Lexington (A. E. Scott). Not common. June-July.

R. triflorus, Rich. DWARF RASPBERRY.
Widely spread, but not very common. May-June.

R. strigosus, Michx. RED RASPBERRY.
Common. Occasionally found with variegated leaves. June.

R. occidentalis, L. BLACK RASPBERRY. THIMBLEBERRY.
Rather common. June.

R. villosus, Ait. HIGH BLACKBERRY.
Common. The forms known as var. **frondosus** and **humifusus** occasional. May-July.

R. Canadensis, L. LOW BLACKBERRY. DEWBERRY.
Common. May-June.

R. hispidus, L. RUNNING SWAMP BLACKBERRY.
Common. June. The form known as var. **setosus,** Bigel., common.

ROSA, Tourn.

R. Carolina, L. SWAMP ROSE.
Common. June-July.

R. lucida, Ehrh. DWARF WILD ROSE.
Very common. May-June.

R. humilis, Marsh. (R. lucida, Man., in part).
Arlington (Mrs. P. D. Richards); Weston (L. L. Dame); Concord (Walter Deane); Ashby (Dr. C. W. Swan). June.
"In dry soil and on rocky slopes and mountain sides. Stems usually low (1 to 3 ft.), and more slender, less leafy, with straight slender spines, spreading or sometimes reflexed; stipules narrow, rarely somewhat dilated; leaflets as in the last, but usually thinner and paler, glabrous or usually more or less pubescent, especially beneath, and also the rachis (often prickly); flowers very often solitary, the outer sepals always more or less lobed, often pinnately so; fruit as in the preceding." Watson in Revision of Roses of N. A., Proc. of Amer. Acad., Vol. XX, 1885.

R. nitida, Willd. (R. lucida, Ehrh., Man., in part.)
Medford, Lexington, Woburn, Bedford, et al. Occasional in low grounds. June.
"Usually low, nearly or quite glabrous throughout, the straight slender spines often scarcely stouter than the prickles which cover the stem and branches more or less thickly; stipules usually dilated; leaflets bright green and shining, usually narrowly oblong and acute at each end, sometimes broader and obtuse, small (the terminal $\frac{1}{2}$ to $1\frac{1}{4}$ in. long); flowers usually solitary (rarely 2 or 3), bright red ($1\frac{1}{2}$ to $2\frac{1}{2}$ in. broad), the slender pedicels, receptacle, and calyx densely hispid or glandular-prickly; sepals entire; fruit globose, 4 or 5 lines broad." Watson in Revision of Roses of N. A., Proc. of Amer. Acad., Vol. XX, 1885.

R. RUBIGINOSA, L. SWEET-BRIER.
Widely distributed, but not very common. June-July. Nat. from Eu.

R. MICRANTHA, Smith. SMALLER-FLOWERED SWEET-BRIER.
Occasional. June-July. Nat. from Eu. Too near R. rubiginosa, and not separated from it by some botanists.

R. cinnamomea, L.
Persistent in old gardens, and occasionally spontaneous. For description, see Wood's Bot. & Fl.

CRATÆGUS, L.

C. Oxyacantha, L. ENGLISH HAWTHORN.
Occasionally spontaneous, but hardly naturalized. May-June. Adv. from Eu.

C. coccinea, L. SCARLET-FRUITED THORN.
Widely distributed; infrequent eastward, but rather common in other sections of the county. May.

C. tomentosa, L., var. **pyrifolia,** Gray.
Cambridge (*fide* specimen in the Gray Herb.) May.

C. tomentosa, L., var. **punctata,** Gray.
Shrubs closely approaching this form have been observed at Medford and Hudson; and there is a specimen from Cambridge " intermediate between **pyrifolia** and **punctata**" in the Gray Herb. May.

C. subvillosa, Schrad. (C. tomentosa, L., var. mollis, Gray, Man.)
Medford (L. L. Dame). Apparently native. Rare. May.

PIRUS, L.

P. COMMUNIS, L. PEAR.
Occasionally spontaneous, but less common than P. malus. May. Nat. from Eu.

P. MALUS, L. APPLE.
Frequently spontaneous in pastures and woods. May. Nat. from Eu.

P. arbutifolia, L., (var. **erythrocarpa,** Gray, Man.) CHOKEBERRY.
Common. May-June.

P. arbutifolia, L., var. **melanocarpa,** Gray.
Common. May-June.

P. Americana, DC. MOUNTAIN ASH.
Ashby and Townsend (L. L. Dame). Not common. June.

AMELANCHIER, Medic.

A. Canadensis, Torr. and Gray, (including var. **Botryapium,** Man.) JUNEBERRY. SHADBUSH.
Common. Apr.-May.

A. Canadensis, Torr. and Gray, var. **oblongifolia,** Gray.
Common. Apr.-May.

SAXIFRAGACEÆ. SAXIFRAGE FAMILY.

RIBES, L.

*R. *Cynosbati*, L. PRICKLY GOOSEBERRY.
Concord, introduced from N. H. by Minot Pratt. May.

R. oxyacanthoides, L. (R. hirtellum, Michx., Man.) WILD GOOSEBERRY.
Common. May.

*R. rotundifolium, Michx.
Reading (W. H. Manning). June.

R. prostratum, L'Her. FETID CURRANT.
Ashby, rather common on and near Mt. Watatic (L. L. Dame). May.

R. floridum, L'Her. WILD BLACK CURRANT.
Generally distributed in the northern and eastern parts of the county, but not very common. May.

R. RUBRUM, L. RED CURRANT.
Spontaneous in several localities. May-June. Nat. from Eu.

PARNASSIA, Tourn.

P. Caroliniana, Michx. GRASS OF PARNASSUS.
Reading (W. H. Manning); Wakefield (Mrs. P. D. Richards); Groton (C. W. Jenks); Framingham, abundant (Rev. J. H. Temple); Ashland (Rev. Thos. Morong). Rather scarce. July-Aug.

SAXIFRAGA, L.

S. Virginiensis, Michx. EARLY SAXIFRAGE.
Very common. Apr.-May.

S. Pennsylvanica, L. SWAMP SAXIFRAGE.
Common. May-June.

MITELLA, Tourn.

*M. diphylla, L. MITREWORT.
Groton (Miss H. E. Haynes); Concord, introduced from Vermont by Minot Pratt. May-June.

TIARELLA, L.

T. cordifolia, L. FALSE MITREWORT. Groton (C. W. Jenks); Concord, introduced from Vermont by Minot Pratt. May-June.

CHRYSOSPLENIUM, Tourn.

C. Americanum, Schw. GOLDEN SAXIFRAGE.
Common. Apr.-May.

CRASSULACEÆ. ORPINE FAMILY.

PENTHORUM, Gronov.

P. sedoides, L. DITCH STONECROP.
Common. July-Sept.

SEDUM, Tourn.

S. ACRE, L. MOSSY STONECROP.
Melrose and Somerville (C. E. Perkins); Reading (Dr. C. W. Swan). Int. from Eu., and sparingly naturalized. June-July.

S. TELEPHIUM, L. LIVE-FOR-EVER. AARON'S ROD.
Roadsides, rather common. July-Aug. Nat. from Eu.

SEMPERVIVUM, L.

S. TECTORUM, L. HOUSELEEK.
Concord (Minot Pratt); Medford and Woburn, persistent and spreading (L. L. Dame). Nat. from Eu. For description, see Wood's Bot. & Fl.

HAMAMELACEÆ. WITCH-HAZEL FAMILY.

HAMAMELIS, L.

H. Virginica, L. WITCH-HAZEL.
Common. Oct.-Dec.

HALORAGEÆ. WATER-MILFOIL FAMILY.

MYRIOPHYLLUM, Vaill.

M. spicatum, L.
Cambridge (Rev. Thos. Morong); Mystic Pond (Wm. Boott). Specimen in the Boott Herb. July-Aug.

M. verticillatum, L.
Fresh Pond, Cambridge (C. E. Perkins). July-Aug.

M. ambiguum, Nutt., (var. natans, of Man.)
Tewksbury (B. D. Greene); Spot Pond (Wm. Boott). Specimen in Boott Herb. July-Aug.

M. ambiguum, Nutt., var. capillaceum, Torr. & Gray.
Townsend and Bedford (Dr. C. W. Swan); Spot Pond (Rev. Thos. Morong); Concord, abundant (Walter Deane); Mystic Pond (Wm. Boott). Growing entirely beneath the surface.

M. ambiguum, Nutt., var. limosum, Torr.
Small ponds in Middlesex Fells (Wm. Boott). Growing in the mud entirely out of water. Possibly this variety, with the one immediately preceding, may be merely forms of the first, dependent on the presence or absence of water and its depth.

M. tenellum, Bigel.
Cambridge and Tewksbury (Bigelow's Fl. Bost.); Natick (Austin Bacon); Silver Lake and Mystic Pond (Wm. Boott); Groton and Westford (Dr. C. W. Swan). July-Aug.

PROSERPINACA, L.

P. palustris, L. MERMAID WEED.
Common. June-July.

TRAPA, L.

T. NATANS, L. WATER CHESTNUT.
Medford and Malden (F. S. Collins); Concord River (C. W. Jenks). Introduced from Europe, and, in the last locality, apparently naturalized.
"Rootstock furnished at the joints with tufts of roots, each tuft forming a pyramidal plume; leaves floating, about 1½ in. broad, rhomboidal, thickish and nerved, bimucronately toothed, subpubescent at the nerves beneath; petiole 2-3 times longer, distended below blade into an oblong intumescence, filled with cellular pith, and acting as a buoy; flowers small, white, submersed, pellucid; peduncles 1-flowered, axillary." Bot. Reg., Vol. 1.

ONAGRACEÆ.   EVENING-PRIMROSE FAMILY.

CIRCÆA, Tourn.

C. Lutetiana, L. ENCHANTER'S NIGHTSHADE.
Common. June-July.

C. alpina, L.
Generally distributed, but not so common as the preceding. July.

EPILOBIUM, L.

E. angustifolium, L. (E. spicatum, Lam.) GREAT WILLOW-HERB.
Very common. July-Aug.

E. palustre, L., var. **lineare,** Gray.
Common. Aug.-Sept.

E. coloratum, Muhl.
Common. July-Sept.

ŒNOTHERA, L.

Œ. biennis, L. EVENING PRIMROSE.
Very common. June-Sept.

Œ. biennis, L., var. **muricata,** Lindl.
Cambridge (F. S. Collins).

Œ. biennis, L. var., **cruciata,** Torr. & Gray.
Chelmsford (Dr. C. W. Swan); Woburn (Mrs. P. D. Richards).
Œ. biennis, L., var. *grandiflora,* Lindl.
Malden, escaped (F. S. Collins).
Œ. pumila, L.
Common. June-July.
Œ. bistorta, Nutt.
Wool-waste, N. Chelmsford, abundant (Rev. W. P. Alcoit). Adv. from Cal.
"Somewhat hirsute, the leaves sometimes appressed pubescent; stems rather stout, decumbent or ascending, a foot or two high; leaves thinner, narrowly lanceolate to ovate, the upper mostly sessile and rounded or cordate at base, all denticulate or dentate; petals 4 to 7 lines long, usually with a dark brown spot at base; capsule 4 to 9 lines long, a line or more wide, attenuate upward; seeds nearly black." Bot. Cal.
Œ. bistorta, Nutt., var. *Veitchiana,* Hook.
Wool-waste, N. Chelmsford, rare (Rev. W. P. Alcott). Adv. from Cal.
"More slender; capsule more elongated and narrowed (1 to 1½ inches long and less than a line broad), attenuate into a narrow beak." Bot. Cal.

LUDWIGIA, L.

L. alternifolia, L. SEED-BOX.
Cambridge (Bigelow's Fl. Bost.); Tewksbury (B. D. Greene). Dr. Swan reports at Lowell a form with fascicled, fusiform roots, a character ascribed by the Man. to L. hirtella, Raf. Scarce. Aug.-Sept.

L. sphærocarpa, Ell.
Tewksbury (B. D. Greene); Waltham (C. E. Perkins); Billerica and Concord (Wm. Boott); Bedford (Walter Deane); Lowell, abundant on the banks of the Concord (Dr. C. W. Swan). Aug.-Sept.

L. polycarpa, Short & Peter.
Waltham List; Winchester, Winter Pond, 1886 (Wm. Boott). Specimen in the Boott. Herb. Aug.-Sept.

L. palustris, Ell. WATER PURSLANE.
Common. July-Oct.

CLARKIA, Pursh.

C. rhomboidea, Dougl.
Wool-waste, N. Chelmsford (Rev. W. P. Alcott; specimen in herb. of). Adv. from Eu. For description, see Wood's Bot. & Fl.

MELASTOMACEÆ. MELASTOMA FAMILY.

RHEXIA, L.

R. Virginica, L. MEADOW BEAUTY.
Not uncommon. July-Aug.

LYTHRACEÆ. LOOSESTRIFE FAMILY.

AMMANNIA, HOUSTON.

A. humilis, Michx.
Winchester,Winter Pond (Dr. C. W. Swan). Rare. July-Sept.

LYTHRUM, L.

L. Hyssopifolia, L. LOOSESTRIFE.
Malden (W. H. Manning; Mrs. C. E. Pease); Medford (Wm. Boott); Arlington (F. S. Collins). Rare. July.
L. alatum, Pursh.
Chelmsford (Miss C. E. Preston). Probably introduced in western wool. June-Aug.
L. Salicaria, L. SPIKED LOOSESTRIFE.
Chelmsford (W. H. Manning); Ashland, rather common (Rev. Thos. Morong); Framingham (Rev. J. H. Temple). Rare northward. July-Aug. Possibly introduced.
L. acutangulum, Lag.
Lowell, a single specimen on the sandy bank of the Merrimac (Dr. C. W. Swan).
"Herbaceous, leaves alternate, linear, lanceolate; pedicels short, erect even in fruit; bractlets acute, very small; petals 6, oblong-ovate; stamens 12." DC. Prodr. III, 82.

NESÆA, Commerson, Juss.

N. verticillata, HBK. SWAMP LOOSESTRIFE.
Common, especially in the northern part of the county. July-Aug.

CACTACEÆ. CACTUS FAMILY.

OPUNTIA, Tourn.

*****O. VULGARIS,** Haworth. PRICKLY PEAR.
N. Reading, bank of the Ipswich river (J. Robinson, Flora of Essex). June-July. Int. from farther south.

CUCURBITACEÆ. GOURD FAMILY.

SICYOS, L.

S. ANGULATUS, L. ONE-SEEDED STAR-CUCUMBER.
N. Reading, Malden, Cambridge, et al. Growing by the roadside or upon rubbish heaps. July-Sept. Nat. from farther west.

ECHINOCYSTIS, Torr. & Gray.

E. LOBATA, Torr. & Gray. WILD BALSAM-APPLE.
Lowell, Concord, Malden, Weston, et al. In localities similar to the preceding. July-Sept. Nat. from farther west.

UMBELLIFERÆ. PARSLEY FAMILY.

HYDROCOTYLE, Tourn.

H. Americana, L. WATER PENNYWORT.
Common. July-Aug.

H. umbellata, L.
Occasional along the banks of Charles River, Martin's, Hammond's, Fresh Ponds, et al. July-Aug.

SANICULA, Tourn.

S. Marilandica, L. BLACK SNAKEROOT.
Not uncommon. June-July.

DAUCUS, Tourn.

D. CAROTA, L. CARROT.
Rather common. July-Sept. Nat. from Eu.

HERACLEUM, L.

H. lanatum, Michx. COW-PARSNIP.
Natick (Austin Bacon); Concord (Minot Pratt); Townsend (Miss H. E. Haynes); Melrose (Bradford Torrey). Rare. June.

PASTINACA, Tourn.

P. SATIVA, L. PARSNIP.
Throughout the county, but nowhere very common. July. Nat. from Eu.

ANGELICA, L.

A. atropurpurea, L. (Archangelica atropurpurea, Hoffm., Man.)
GREAT ANGELICA.
Generally distributed, but scarce. Specimen from Watertown in the Boott Herb. June-July.

ÆTHUSA, L.

Æ. CYNAPIUM. L. FOOL'S PARSLEY.
Roadsides, Medford to Watertown (Wm. Boott). Nat. from Eu. This plant has poisonous qualities, and serious results have sometimes ensued from confounding it with common Parsley, which it somewhat resembles, but from which it may easily be distinguished, when in flower, by its lack of a general involucre, and its long, hanging involucels. July.

LIGUSTICUM, L.

*L. Scoticum, L. SCOTCH LOVAGE.
Cambridge (Bigelow's Fl. Bost.); Watertown (C. E. Perkins). Rare. Aug.

THASPIUM, Nutt.

T. aureum, Nutt. MEADOW PARSNIP.
Common in the northern and western parts of the county. May-June.

ZIZIA, Koch. (Not ZIZIA of Man.)

Z. aurea, Koch. (Thaspium aureum, Nutt., var. apterum Man.)
Dunstable (Dr. C. W. Swan).

BUPLEURUM, Tourn.

B. rotundifolium, L.
Cambridge, one plant in gravel sidewalk (Walter Deane; specimen in herb. of). Adv. from Eu.

CICUTA, L.

C. maculata, L. SPOTTED COWBANE. WATER HEMLOCK.
Common. July-Aug.
C. bulbifera, L.
Common. Aug.-Sept.

SIUM, L.

S. cicutæfolium, Gmel. (S. lineare, Michx., Man.) WATER PARSNIP.
Common. July-Aug.
S. Carsoni, Durand.
Tewksbury (Dr. C. W. Swan); Ashland (Rev. Thos. Morong). Very rare.

CRYPTOTÆNIA, DC.

C. Canadensis, DC. HONEWORT.
Belmont, abundant in 1882 (Walter Deane: specimen in herb. of). June-July.

OSMORRHIZA, Raf.

O. longistylis, DC. SMOOTHER SWEET CICELY.
Watertown, Belmont, Waverly, et al. Not uncommon. May-June.

O. brevistylis, DC. HAIRY SWEET CICELY.
Woods, Concord Turnpike (Bigelow's Fl. Bost.); Cambridge (B. D. Greene); Malden (F. S. Collins). Scarce. May-June.

CONIUM, L.

C. MACULATUM, L. POISON HEMLOCK.
Watertown (C. E. Perkins); Natick (Austin Bacon); Arlington (Wm. Boott); Waltham List. Rare. July-Aug. Nat. from Eu.

CARUM, L.

C. Carui, L. CARAWAY.
Occasional. July. Adv. from Eu.

ARALIACEÆ. GINSENG FAMILY.

ARALIA, Tourn.

A. racemosa, L. SPIKENARD.
Rather scarce, but generally distributed. July.

A. hispida, Vent. BRISTLY SARSAPARILLA.
Rather common. June.

A. nudicaulis, L. SARSAPARILLA.
Common. May-June.

A. quinquefolia, Gray. GINSENG.
Concord, introduced from Vermont, but does not thrive (Minot Pratt). July.

A. trifolia, Decne and Planch. DWARF GINSENG. GROUND NUT.
Generally distributed, but not common. May-June.

CORNACEÆ. DOGWOOD FAMILY

CORNUS, Tourn.

C. Canadensis, L. DWARF CORNEL. BUNCHBERRY.
Common. June.

C. florida, L. FLOWERING DOGWOOD.
Not uncommon. May-June.

C. circinata, L'Her. ROUND-LEAVED CORNEL.
Rather common. June.

C. sericea, L. SILKY CORNEL.
Common. June.

C. stolonifera, Michx. RED-OSIER DOGWOOD.
Concord (Minot Pratt); Cambridge, under C. alba, Lam. (Bigelow's Fl. Bost.); Woburn (L. L. Dame); Reading (W. H. Manning); et al. Rather scarce. June.
C. paniculata, L'Her. PANICLED CORNEL.
Common. June.
C. alternifolia, L. ALTERNATE-LEAVED CORNEL.
Common. May-June.

NYSSA, L.

N. sylvatica, Marsh. (N. multiflora, Wang., Man.) TUPELO. PEPPERIDGE.
Generally distributed, but scarce. May-June.

CAPRIFOLIACEÆ. HONEYSUCKLE FAMILY.

LINNÆA, Gronov.

L. borealis, Gronov. TWIN-FLOWER.
Not reported in the eastern part of the county, and infrequent in the other sections. June.

LONICERA, L.

L. sempervirens, Ait. TRUMPET HONEYSUCKLE.
Marlboro (Mrs. A. M. Staples); Medford (L. L. Dame). Sometimes escaping from cultivation; but abundant, remote from dwellings, and apparently native in the Medford locality. June-July.
L. Tatarica, L.
Occasionally spontaneous. May-June.
***L. hirsuta,** Eaton. HAIRY HONEYSUCKLE.
Sudbury (Emerson's Mass. Trees and Shrubs, 1846). June-July.
L. ciliata, Muhl. FLY HONEYSUCKLE.
Townsend (Miss H. E. Haynes); Framingham (Rev. J. H. Temple). Rare.

DIERVILLA, Tourn.

D. trifida, Mœnch. BUSH HONEYSUCKLE.
Common. June-July.

TRIOSTEUM, L.

T. perfoliatum, L. HORSE-GENTIAN.
Eastern and southern parts of the county. Not very common. June-July.

SAMBUCUS, Tourn.

S. Canadensis, L. ELDER.
Common. July.

S. racemosa, L. (S. pubens, Michx., Man.) RED-BERRIED ELDER.
Ashby, not uncommon (L. L. Dame); occasional in other parts of the county. In the Concord station the plant was introduced from Wachuset by Minot Pratt. May.

VIBURNUM, L.

V. cassinoides, L. (V. nudum, var. cassinoides, Man.) WITHE-ROD.
Not very common in the eastern part of the county; abundant at Ashby and vicinity. June.

V. Lentago, L. SWEET VIBURNUM. SHEEP-BERRY.
Rather common. May-June.

V. dentatum, L. ARROW-WOOD.
Common. June.

V. acerifolium, L. MAPLE-LEAVED VIBURNUM.
Rather common. June.

V. Opulus, L. CRANBERRY TREE.
Groton (C. W. Jenks); Weston (L. L. Dame); Pepperell (Dr. C. W. Swan); Concord, introduced from Vermont by Minot Pratt. June.

V. lantanoides, Michx. HOBBLE-BUSH.
Ashby, not uncommon (W. H. Manning). Not authoritatively reported elsewhere. May-June.

RUBIACEÆ. MADDER FAMILY.

GALIUM, L.

G. Aparine, L. CLEAVERS. GOOSE-GRASS.
Lowell, "dumps" (Dr. C. W. Swan); Medford (C. E. Perkins); Waltham List; Townsend (Miss H. E. Haynes). Rare. May.

G. Mollugo, L.
McLean asylum grounds, Somerville (C. E. Perkins). Adv. from Eu.

G. asprellum, Michx. ROUGH BEDSTRAW.
Common. July-Aug.

G. trifidum, L., (including var. **tinctorium,** of the Manual.) SMALL BEDSTRAW.
Common. June-July.

G. triflorum, Michx. SWEET-SCENTED BEDSTRAW.
Generally distributed, but not very common. July.

G. pilosum, Ait.
Lowell (Dr. C. W. Swan); Malden (H. A. Young); Concord (Minot Pratt); Waltham (Wm. Boott). Rare. June-Aug.

G. circæzans, Michx. WILD LIQUORICE.
Rather common. June-Aug.

G. lanceolatum, Torr. WILD LIQUORICE.
Melrose, Malden, Stoneham, et al. Not very common. June-July.

G. VERUM, L. YELLOW BEDSTRAW.
Arlington (Wm. Boott); Natick (Austin Bacon). Rare. July. Nat. from Eu.

CEPHALANTHUS, L.

C. occidentalis, L. BUTTON-BUSH.
Common. July-Aug.

MITCHELLA, L.

M. repens, L. PARTRIDGE-BERRY.
Very common. A variety with white fruit has been reported at Concord by A. W. Hosmer. June-July.

HOUSTONIA, L.

H. purpurea, L., var. **longifolia,** Gray.
Marlboro (Mrs. A. M. Staples); common in Woburn and Lexington; less common in the adjacent towns, and rare in other sections of the county. June-Aug.

H. cœrulea, L. BLUETS. INNOCENCE.
Very common. May-Aug.

COMPOSITÆ. COMPOSITE FAMILY.

VERNONIA, Schreb.

V. Noveboracensis, Willd. IRON-WEED.
Chiefly in the central and southern portions of the county. Not common. Aug.

V. fasciculata, Michx.
West Medford, a flourishing patch in 1886, reappearing in 1887 (Mrs. P. D. Richards). Adv. from the West. Aug.

LIATRIS, Schreb.

L. scariosa, Willd.
Rather common. A form with flowers pure white reported at Medford (Mrs. P. D. Richards). Aug.-Sept.

*L. spicata, Willd.
Framingham, rare, probably introduced (Rev. J. H. Temple); Concord, introduced from Ill. by Minot Pratt.

EUPATORIUM, Tourn.

E. purpureum, L. TRUMPET WEED. QUEEN OF THE MEADOW.
Common. Aug.-Sept.

E. teucrifolium, Willd.
Medford (H. L. Moody; G. E. Davenport); Cambridgeport, as
E. verbenæfolium (Bigelow's Fl. Bost.); Framingham (Rev. J. H.
Temple); Melrose (Rev. Thos. Morong). Rare. Aug.-Sept.

E. rotundifolium, L.
Swain's Pond, Melrose (H. L. Moody; C. E. Perkins). Rare.
Aug.-Sept.

E. rotundifolium, × teucrifolium, (*Fide* Asa Gray.) Lowell
and Chelmsford (Dr. C. W. Swan). Aug.-Sept.

E. rotundifolium, L., var. **ovatum,** Torr. (E. pubescens,
Muhl., Man.)
Malden (H. L. Moody); Arlington and Sudbury, under E. ovatum
(Bigelow's Fl. Bost.); Natick (Austin Bacon); Waltham List.
Not common. Aug.-Sept.

E. sessilifolium, L. UPLAND BONESET.
Malden (F. S. Collins); Woburn, abundant, Medford and Hudson
(L. L. Dame); Concord, rather rare (Minot Pratt); Waltham List.
Not generally common. Aug.-Sept.

E. perfoliatum, L. THOROUGHWORT. BONESET.
Common. Aug.-Sept.

E. perfoliatum, L., var. **truncatum,** Gray.
Dracut (Dr. C. W. Swan). Aug.-Sept.
"With the upper or even all of the leaves disjoined and truncate at
the base; some of them alternate." Syn. Fl. N. A.

E. ageratoides, L. WHITE SNAKE-ROOT.
Tewksbury and Dracut (Dr. C. W. Swan); Malden (H. L. Moody;
F. S. Collins); Waltham List. The Concord plant was introduced
from Wachuset by Minot Pratt. Scarce. Aug.-Sept.

E. aromaticum, L.
Malden (H. L. Moody); Stoneham (F. S. Collins). Rare. Aug.-
Sept.

MIKANIA, Willd.

M. scandens, Willd. CLIMBING HEMP-WEED.
Lowell, Framingham, Medford, et al. Scarce. July-Sept.

TUSSILAGO, Tourn.

T. FARFARA, L. COLTSFOOT.
Somerville (C. E. Perkins); Marlboro (Mrs. A. M. Staples); formerly at Groton, now extinct (C. W. Jenks). The Concord plant
was introduced from Vt. by Minot Pratt. Scarce. Apr. Nat.
from Eu.

SERICOCARPUS, Nees.
S. solidagineus, Nees. WHITE-TOPPED ASTER.
Medford (C. E. Perkins); Natick (Austin Bacon); Bedford (C. W. Jenks); Chelmsford (Dr. C. W. Swan). Not very common. July-Aug.
S. conyzoides, Nees.
Common. July.

ASTER L.

A. corymbosus, Ait.
Rather common. July-Aug.
A. macrophyllus, L.
Rather common. July-Aug.
A. radula, Ait.
Not uncommon. Aug.
A. patens, Ait.
Common. Aug.-Oct.
A. lævis, L.
Typical form and vars. **lævigatus** and **cyaneus**, Man. Common. Aug.-Oct.
A. undulatus, L.
Common. Aug.-Oct.
A. cordifolius, L.
Not uncommon. Aug.-Oct.
A. ericoides, L.
Westford (Misses Fletcher and Hodgman); Waltham List; Medford (Mrs. P. D. Richards); Bedford (C. E. Faxon). Sept.-Oct.
A. multiflorus, Ait.
Rather common. Aug.-Oct.
A. dumosus, L.
Not very common. Aug.-Sept.
A. Tradescanti, L., (not of Man.)
Cambridge (L. H. Bailey, Jr.) Specimen in Gray Herb. Aug.-Sept.
A. vimineus, Lam. (A Tradescanti and var. fragilis, Man.)
Common. Aug.-Oct.
A. diffusus, Ait. (A. miser, Man.)
Rather common. Aug.-Sept.
A. paniculatus, Lam. (A. simplex, Willd., Man.)
Lowell, Tyngsborough and Marlboro (Dr. C. W. Swan); Malden (F. S. Collins); Waltham List; Ashland (Rev. Thos. Morong). Scarce. Aug.-Sept.
A. salicifolius, Ait. (A. carneus, Nees., in part, Man.)
Lowell (Dr. C. W. Swan); Malden (H. L. Moody); "woods on Concord Turnpike; (Bigelow's Fl. Bost.) Scarce.

A. Novi Belgii, L. (A. longifolius, Lam., Man.)
Widely distributed; not uncommon. Occasional with white flowers. Aug.-Oct.

A. Novi Belgii, L., var. **litoreus,** Gray.
Not uncommon near salt marshes. Sept.-Oct.
" Stems rigid, low, or sometimes 3 or 4 feet high, and then paniculately much branched, very leafy; leaves thickish and firm, very smooth, (rarely upper face somewhat scabrous), oblong to lanceolate, upper partly clasping and sometimes auriculate; bracts of the involucre loosely imbricated in several ranks, outer commonly spatulate, all but innermost with broadish or obtuse herbaceous and mostly thickish tips." Syn. Fl., N. A.

A. puniceus, L.
Common. Aug.-Oct.

A. puniceus, L., var. **lucidulus,** Gray, (var. vimineus, Man.)
Malden (H. L. Moody); Medford (L. L. Dame). Sept.-Oct.

A. amethystinus, Nutt.
Westford (Misses Fletcher and Hodgman); Cambridge (H. L. Moody); Arlington (A. E. Verrill); Belmont (Wm. Boott). Scarce. Aug.-Sept.

A. Novæ Angliæ, L. New England Aster.
Rather common. Aug.-Oct.

A. acuminatus, Michx.
Common. Aug.-Sept.

A. nemoralis, Ait.
Tewksbury (Wm. Boott); Dracut (Dr. C. W. Swan); Long Pond. Melrose (H. L. Moody). Rare. Aug.-Sept.

A. tenuifolius, L. (A. flexuosus, Nutt., Man).
Borders of Mystic Pond, Medford (Rev. Thos. Morong; specimen in herb. of). Rare.

A. subulatus, Michx. (A. linifolius, Torr. & Gray, Man.)
Common along salt marshes. Aug.-Sept.

A. linariifolius, L. (Diplopappus linariifolius, Hook., Man.)
Common. A form with white flowers not uncommon. Aug.-Sept.

A. umbellatus, Mill. (Diplopappus umbellatus, Torr. & Gray, Man.)
Common. Aug.-Oct.

A. infirmus, Michx. (Diplopappus cornifolius, Torr. & Gray, Man.)
Concord (F. S. Collins); Acton (Dr. C. W. Swan). Rare. Sept.

ERIGERON, L.

E. Canadensis, L. HORSEWEED.
Common. July-Oct.

E. bellidifolius, Muhl. ROBIN'S PLANTAIN.
Common. May-June.

E. Philadelphicus, L. FLEABANE.
Groton (C. W. Jenks); No. Reading (Dr. C. W. Swan); Townsend (Miss H. E. Haynes). Rare. June-July.

E. annuus, Pers. DAISY FLEABANE.
Common. June-Sept.

E. strigosus, Muhl. DAISY FLEABANE.
Common. The form known as var. **discoideus,** Robbins, occasional. June-Aug.

BOLTONIA, L'Her.

B. asteroides, L'Her. (B. glastifolia, L'Her., Man.)
Concord, int. from Penn. by Minot Pratt.

BELLIS, Tourn.

B. perennis, L. ENGLISH DAISY.
Malden, near B. & M. R. R., May, 1886 (F. S. Collins). Adv. from Eu. For description, see Wood's Bot. and Fl.

SOLIDAGO, L.

S. squarrosa, Muhl.
Burlington (Miss M. E. Carter); Concord, introduced from Vt. by Minot Pratt. The Burlington plant is unquestionably native, though the locality is limited, and fortunately not easily found by over-enthusiastic collectors. Aug.-Sept.

S. bicolor, L.
Common. Aug.-Sept.

S. bicolor, L., var. **concolor,** Gray.
Woburn (Dr. C. W. Swan); Weston (Walter E. Coburn). Rare. Aug.-Sept.

S. latifolia, L.
Not very common. Aug.-Sept.

S. cæsia, L.
Common. Aug.-Oct.

S. puberula, Nutt.
Not very common. Aug.-Sept.

***S. uliginosa,** Nutt. (S. stricta, Ait., Man.)
Waltham List; Concord, peat bogs (Minot Pratt). Rare. Aug.-Sept.

S. speciosa, Nutt.
Westford (Misses Fletcher and Hodgman); Malden (H. L. Moody); Waltham and Arlington (Wm. Boott); Winchester (L. L. Dame). Rare. Sept.-Oct.

S. sempervirens, Michx.
Common along the salt marshes. Aug.-Sept.

S. neglecta, Torr. & Gray.
Common. Aug.-Sept.

S. neglecta, Torr. & Gray, var. **linoides,** Gray (S. linoides, Soland., Man.)
Malden, 1887 (E. H. Hitchings). Sept.

S. juncea, Ait. (S. arguta, of Man.)
Common. July-Aug.

S. arguta, Ait. (S. Muhlenbergii, Torr. & Gray., Man.)
Medford, Lowell, Tewksbury, et al.; apparently more common in the northern portion of the county. July-Sept.

S. rugosa, Mill. (S. altissima, L., Man.)
Common. Aug.-Sept.

S. ulmifolia, Muhl.
Dracut (Dr. C. W. Swan); Malden (H. L. Moody); Medford (Wm. Boott); Melrose and Woburn (Mrs. P. D. Richards). Not very common. Aug.-Sept.

S. odora, Ait. SWEET GOLDEN-ROD.
Widely distributed, but not very common. July-Aug.

S. nemoralis, Ait.
Common. Aug.-Oct.

S. Canadensis, L.
Common. July-Sept. A peculiar form near Fresh Pond, with the stem perfectly smooth up to the inflorescence; leaves scabrous above, ciliate, hairy on the veins beneath, (Walter Deane).

S. Canadensis, L., var. **procera,** Torr. & Gray., Malden, (H. L. Moody).

S. serotina, Ait. (S. gigantea, of Man.)
Not very common. July-Aug.

S. serotina, Ait., var. **gigantea,** Gray. (S. serotina of Man.)
Medford (L. L. Dame); Malden (H. L. Moody); Stoneham (W. H. Manning); Waltham List; Concord (Minot Pratt). Not uncommon. Aug.-Sept.

S. lanceolata, L.
Common. Aug.

S. tenuifolia, Pursh.
Silver Lake, Wilmington (J. R. Churchill); Tewksbury (Wm. Boott); the Concord plant was introduced from Weymouth by Minot Pratt. July-Sept.

GRINDELIA, Willd.

G. robusta, Nutt. GUM-PLANT.
Lowell, "dumps" (Dr. C. W. Swan). Adv. from Cal. For description see Gray, Syn. Fl. N. A.

INULA, L.

I. Helenium, L. Elecampane.
Generally distributed, but not very common. July-Aug. Nat. from Eu.

PLUCHEA, Cass.

P. camphorata, DC. Salt-marsh Fleabane.
Common along salt marshes. Aug.

SILPHIUM, L.

*S. laciniatum, L. Rosin-weed. Compass-plant.
Concord, int. from Ill. by Minot Pratt. July.
*S. terebinthinaceum, L. Prairie Dock.
Concord, int. from Ill. by Minot Pratt. July-Aug.

PARTHENIUM, L.

P. Hysterophorus, L.
Lowell, "dumps" (Dr. C. W. Swan). Sept. Adv. from the South. For description, see Wood's Bot. & Fl.

IVA, L.

I. frutescens, L.
Common along salt marshes and tidal streams. Aug.

AMBROSIA, Tourn.

A. artemisiæfolia, L. Roman Wormwood. Ragweed.
Very common. A form with fertile spikes from Malden (F. S. Collins). July-Sept.

XANTHIUM, Tourn.

X. strumarium, L. Cocklebur.
N. Chelmsford, wool-waste (Rev. W. P. Alcott; specimen in herb. of). Aug. Adv. from Eu.

X. Canadense, Mill., var. echinatum, Gray.
Malden and Medford (F. S. Collins); Westford, near woollen-mill (Dr. C. W. Swan). More common near salt marshes. Aug.
X. spinosum, L. Spiny Clotbur.
Chelmsford, Lowell, Dracut and Westford (Dr. C. W. Swan). Aug. Int. in wool from Trop. Am.

ECHINACEA, Moench.

*E. purpurea, Moench. Purple Cone-flower.
Centralville, Lowell (Dr. C. W. Swan); Waltham List. July. Adv. from the West.

RUDBECKIA, L.

R. laciniata, L.
Tewksbury, Lowell, Chelmsford, et al. Not very common. July-Aug.

R. hirta, L. CONE-FLOWER.
Common. July-Aug.

HELIANTHUS, L.

H. annuus, L. COMMON SUNFLOWER.
Occasionally spontaneous. Aug.-Sept. Int. from Trop. Am.

H. strumosus, L.
Widely distributed. Aug.-Sept.

H. divaricatus, L.
Common. Aug.-Sept.

H. decapetalus, L.
Lowell and Tewksbury (Dr. C. W. Swan); Medford (Mrs. P. D. Richards); Malden (F. S. Collins). Not common. Aug.-Sept.

H. petiolaris, Nutt.
Westford, woollen mill (Dr. C. W. Swan). Sept. Adv. from the West. For description, see Gray, Syn. Fl. N. A.

H. TUBEROSUS, L. JERUSALEM ARTICHOKE.
Malden (R. Frohock); Melrose (H. L. Moody); Somerville (F. S. Collins); Concord (Minot Pratt). Scarce. Sept. Nat. from the South.

COREOPSIS, L.

C. rosea, Nutt.
Winchester and Woburn (Wm. Boott). Abundant about Winter and Round Ponds, but nowhere else in the county. Aug.-Sept.

C. aristosa, Michx.
Lowell, "dump" (Dr. C. W. Swan). Aug. Adv. from the West.

C. trichosperma, Michx. TICKSEED SUNFLOWER.
Malden and Melrose (F. S. Collins); Fresh Pond (Bigelow's Fl. Bost.); Wakefield (Mrs. P. D. Richards); Concord (Minot Pratt). Abundant where it occurs at all. Sept.-Oct.

BIDENS, Tourn.

B. frondosa, L. BEGGAR-TICKS.
Common. Aug.-Sept.

B. connata, Muhl. SWAMP BEGGAR-TICKS.
Not reported from the northern and western part of the county, but common elsewhere. Aug.-Sept.

B. cernua, L. SMALLER BUR-MARIGOLD.
Widely distributed, but rather scarce. Aug.-Sept.

B. chrysanthemoides, Michx. LARGER BUR-MARIGOLD.
Natick (Austin Bacon); Acton (Dr. C. W. Swan); Bedford and Billerica (C. W. Jenks); Concord (Minot Pratt). Not very common, and not reported in the northern part of the county. Aug.-Sept.

B. Beckii, Torr. WATER MARIGOLD.
Lowell, Bedford, Woburn, Ashland, et al. Not very common. Aug.-Sept.

B. bipinnata, L. SPANISH NEEDLES.
Lowell, "dump" (Dr. C. W. Swan). Aug.-Sept. Adv. from farther west.

HEMIZONIA, DC.

H. fasciculata, Torr. & Gray, var. *ramosissima,* Gray.
N. Chelmsford, wool-waste (Rev. W. P. Alcott). Adv. from Cal. For description, see Gray, Syn. Fl. N. A.

LAYIA, Hook. & Arn.

L. platyglossa, Gray. TIDY-TIPS.
N. Chelmsford, wool-waste (Rev. W. P. Alcott; specimen in herb. of). Adv. from Cal. For description, see Gray, Syn. Fl. N. A.

BAERIA, Fisch. & Meyer.

B. gracilis, Gray, var. *tenerrima,* Gray.
N. Chelmsford, wool-waste (Rev. W. P. Alcott; specimen in herb. of). Adv. from Cal. For description, see Gray, Syn. Fl. N. A.

B. uliginosa, Gray.
N. Chelmsford, wool-waste (Rev. W. P. Alcott; specimen in herb. of). Adv. from Cal. For description, see Gray, Syn. Fl. N. A.

CHÆNACTIS, DC.

C. glabriuscula, DC.
N. Chelmsford, wool-waste (Rev. W. P. Alcott). Adv. from Cal. For description, see Gray, Syn. Fl. N. A.

HELENIUM, L.

*H. autumnale, L. SNEEZE-WEED.
Concord, int. from Ill. by Minot Pratt. Sept.

H. tenuifolium, Nutt.
Lowell, "dump" (Dr. C. W. Swan); Malden, cotton-waste, Goulding's Mill (F. S. Collins). Adv. from the South. For description, see Woods Bot. & Fl.

GALINSOGA, Ruiz. & Pav.

*G. PARVIFLORA, Cav.
Cambridge, waste places (Man. 2d ed.) Locally nat. from So. Am.

ANTHEMIS, L.

A. COTULA, L. (Maruta Cotula, DC., Man.) MAY-WEED.
Very common. July-Sept. Nat. from Eu.

A. *arvensis*, L. CORN CHAMOMILE.
Lowell (Dr. C. W. Swan). June-July. Adv. from Eu.

ACHILLEA, Vaill.

A. Millefolium, L. YARROW.
Very common. Form with pink flowers not uncommon. July-Sept.

CHRYSANTHEMUM, Tourn., L.

C. LEUCANTHEMUM, L. (Leucanthemum vulgare, Lam., Man.) OX-EYE DAISY. WHITE-WEED.
Very common. June-July. Nat. from Eu.

C. *Parthenium*, Pers. (Leucanthemum Parthenium, Godron, Man.) FEVERFEW.
Dracut, roadside (Dr. C. W. Swan); Malden (F. S. Collins); Waltham List. Escaped from gardens. July-Aug. Int. from Eu.

C. *Balsamita*, L., var. *tanacetoides*, Boiss. COSTMARY. MINT GERANIUM. MINT. GOOSE-TONGUE.
Persistent in old gardens and occasional by the roadside. Aug.-Sept. Adv. from Asia.
"A rayless or discoid form, known by its sweet-scented herbage, barely serrate oblong leaves, and yellowish flowers; when the rays appear they are white." Syn. Fl. N. A.

MATRICARIA, Tourn.

M. *discoidea*, DC.
N. Chelmsford, wool-waste (Rev. W. P. Alcott; specimen in herb. of). Adv. from Cal.

TANACETUM, Tourn.

T. VULGARE, L. TANSY.
Common by the roadside; including var. CRISPUM of the Man., which is not uncommon. July-Aug. Nat. from Eu.

ARTEMISIA, Tourn., L.

A. VULGARIS, L. MUGWORT.
Lowell and Carlisle (Dr. C. W. Swan); Malden and Somerville (F. S. Collins); Marlboro (Mrs. A. M. Staples). Aug.-Sept. Nat. from Eu.

A. biennis, Willd. BIENNIAL WORMWOOD
Cambridge (C. E. Perkins); Billerica (Dr. C. W. Swan). Aug.-Sept. Adv. from the West, and tending towards naturalization.

A. ABSINTHIUM, L. WORMWOOD.
Cambridge (C. E. Perkins); Medford (G. E. Davenport); Malden (F. S. Collins); Groton (Dr. C. W. Swan). Aug. Nat. from Eu.

A. STELLERIANA, Bess.
Tewksbury, a garden escape (Dr. C. W. Swan); Cambridge, growing in the sand on the bank of the Charles (Walter Deane; specimen in herb. of). Int. from the north-west coast, and locally naturalized. "A foot or two high from a creeping lignescent base, robust, densely white-tomentose, the tomentum of the stem cottony; leaves obovate or spatulate in outline, sinuately or incisely pinnatifid; lobes obtuse; corolla glabrous; akenes a line and a half long, oblong, not contracted at summit; the coat utricular." Syn. Fl. N. A.

GNAPHALIUM, L.

G. decurrens, Ives. EVERLASTING.
Dunstable and No. Acton (L. L. Dame); Medford and Ashby (Dr. C. W. Swan); Framingham (Rev. J. H. Temple); Woburn and Medford (Wm. Boott). Widely distributed, but nowhere common. July-Sept.

G. polycephalum, Michx. EVERLASTING.
Common. Aug.-Sept.

G. uliginosum, L. CUDWEED.
Common. July-Sept.

G. purpureum, L. PURPLISH CUDWEED.
Lowell, "dump" (Dr. C. W. Swan). June.

ANAPHALIS, DC.

A. margaritacea, Benth. & Hook. (Antennaria margaritacea, R. Br., Man.) PEARLY EVERLASTING.
Common. Aug.-Sept.

ANTENNARIA, Gaertn.

A. plantaginifolia, Hook. LADIES' TOBACCO.
Very common. April-May.

ERECHTITES, Raf.

E. hieracifolia, Raf. FIREWEED.
Very common. Aug.-Sept.

SENECIO, Tourn.

S. VULGARIS, L. GROUNDSEL.
Somerville (C. E. Perkins); Marlboro (Mrs. A. M. Staples); Lowell (Dr. C. W. Swan); Medford and Malden (F. S. Collins). Not common save in the towns near Boston. July-Oct. Nat. from Eu.

S. viscosus, L.
Lowell, waste ground (Dr. F. Nickerson); Aug.-Oct. Adv. from Eu.
"Coarser than the preceding, viscid-pubescent, strong-scented; leaves once or twice pinnatifid; heads rather larger, more pedunculate; involucre sparingly and slenderly bracteolate at base, its bracts not black-tipped; rays with inconspicuous ligule; akenes glabrous." Syn. Fl. N. A.

S. aureus, L. GOLDEN RAGWORT.
Common. May-June.

S. aureus, L., var. obovatus, Torr. & Gray, Melrose (F. S. Collins).

S. aureus, L., var. Balsamitæ, Torr. & Gray, Stoneham; (Bigelow's Fl. Bost.); Woburn (C. E. Perkins).

CENTAUREA, L.

C. Cyanus, L. BACHELOR'S BUTTON. BLUEBOTTLE.
Escaped from gardens. July-Aug. Int. from Eu.

C. NIGRA, L. KNAPWEED.
Chelmsford (Dr. C. W. Swan); Acton (L. L. Dame); Billerica (C. W. Jenks); and common for many years in the towns near Boston. Forms with white and pink flowers not uncommon; occasional forms with marginal flowers much enlarged. Aug.-Sept. Nat. from Eu.

C. Jacea, L.
Spreading in a field in E. Lexington (Miss Eliza Wellington). Adv. from Eu.
"Heads usually larger (than nigra); brownish appendages of the involucral bracts merely lacerate; marginal flowers neutral and with enlarged palmate corollas, forming conspicuous false rays; otherwise like nigra." Syn. Fl. N. A.

C. leucophæa, Jord.
What may be this species was collected at Westford, near the woollen mills, by Dr. C. W. Swan. The plants are about a foot high, with numerous pinnatifid radical leaves, and stem leaves similar but smaller and less divided, the uppermost short and linear. The heads are about the size of those of C. nigra, rose color, the scales of the involucre conspicuously veined, with darker, ciliate tips.

C. benedicta, L. (Cnicus benedictus, L., Man.) BLESSED THISTLE.
Cambridge, waste-heap, a few plants, 1885 (Rev. Thos. Morong; specimen in herb. of). Adv. from Eu.

CNICUS, Tourn., L.

C. LANCEOLATUS, Hoffm. (Cirsium lanceolatum, Scop., Man.) COMMON THISTLE.
Common. July-Sept. Nat. from Eu.

C. altissimus, Willd., var. **discolor,** Gray. (Cirsium discolor, Spreng., Man.)
Malden (F. S. Collins). Scarce. Aug.-Sept.

C. muticus, Pursh. (Cirsium muticum, Michx., Man.) SWAMP THISTLE.
Not very common. Aug.-Sept.

C. pumilus, Torr. (Cirsium pumilum, Spreng., Man.) PASTURE THISTLE.
Common; a white flowered form occasional. July-Aug.

C. horridulus, Pursh. (Cirsium horridulum, Michx., Man.) YELLOW THISTLE.
Medford and Everett (F. S. Collins). Rather scarce, and found only near salt water. June-July.

C. ARVENSIS, Hoffm. (Cirsium arvense, Scop., Man,) CANADA THISTLE.
Very common; a white flowered form not uncommon. July-Sept. Nat. from Eu.

ONOPORDON, Vaill.

O. ACANTHIUM, L. COTTON THISTLE.
Watertown (Wm. Boott); Tewksbury (C. J. Sprague); Chelmsford (Dr. C. W. Swan); Somerville (C. E. Perkins); Concord (Minot Pratt). Rare. July. Nat. from Eu.

ARCTIUM, L.

A. *Lappa*, L., var. *majus*, Gray. (Lappa officinalis, All., var. major, Man.) LARGER BURDOCK.
Westford, near woollen mill (Dr. C. W. Swan); Malden (F. S. Collins). Aug.-Sept. Adv. from Eu.

A. *Lappa*, L., var. *tomentosum*, Gray.
Cambridge, waste-heap (Rev. Thos. Morong). Adv. from Eu.

A. LAPPA, L., VAR. MINUS, Gray. BURDOCK.
Very common; a form with white flowers occasional; also a form with laciniate leaves. Nat. from Eu.

LAMPSANA, Tourn.

L. COMMUNIS, L. NIPPLEWORT.
Medford (C. E. Perkins); Newton (C. J. Sprague). Nat. from Eu.

CICHORIUM, Tourn.

C. INTYBUS, L. SUCCORY. CHICORY.
Common, especially near Boston. July-Sept. Nat. from Eu.

KRIGIA, Schreb.

K. Virginica, Willd. DWARF DANDELION.
Common. May-June.

HYPOCHÆRIS, L.

"Involucre imbricate; achenium attenuate into a long beak, or slightly attenuate and almost without beak; pappus plumose; receptacle chaffy, chaff deciduous. Flowers yellow or golden." Koch, Syn. Fl. Germ.

H. glabra, L.
N. Chelmsford, wool-waste (Rev. W. P. Alcott; specimen in herb. of). Adv. from Cal., but a native of Eu.
"Stem branching, smooth, leafless; ray florets as long as the involucre; central achenia long-beaked, the marginal without beak." Koch, Syn. Fl. Germ.

MICROSERIS, Don.

M. Douglasii, Gray.
N. Chelmsford, wool-waste (Rev. W. P. Alcott; specimen in herb. of).
"A very rare species, having been collected only by Douglas himself, over forty years ago," Watson *in lit.* 1879. Adv. from the Pacific coast.
For description, see Gray, Syn. Fl. N. A.

LEONTODON, L., Juss.

L. AUTUMNALIS, L. FALL DANDELION.
Very common. Aug.-Oct. Nat. from Eu.

HIERACIUM, Tourn.

H. Canadense, Michx. CANADA HAWKWEED.
Rather common. Aug.-Sept.
H. scabrum, Michx. ROUGH HAWKWEED.
Common. Aug.
H. venosum, L. RATTLESNAKE-WEED.
Common. July-Aug.
H. paniculatum, L. PANICLED HAWKWEED.
Rather common. Aug.-Sept.
H. aurantiacum, L.
Reading, et al.; occasionally spontaneous. June. Adv. from Eu.
For description, see Gray, Syn. Fl. N. A.

CREPIS, L.

"Annuals or perennials, with soft, white pappus and narrow-necked or beaked akenes (some truncate or merely tapering upwards); leaves entire or inclined to be pinnatifid; flowers all yellow." Coulter, R. M. Bot.

C. fœtida, L.
Westford, woollen-mill yard (Dr. C. W. Swan). Adv. from Eu.
"Stem erect, leafy, branching, hairy; leaves hairy, runcinate-pinnatifid, the uppermost lanceolate, deeply incised at the base; peduncles drooping before flowering; beaks of the marginal achenia shorter than the involucre, those of the interior longer; the entire involucre downy and villous; hairs simple and glandular; bracts of the involucre lanceolate, acute." Koch, Syn. Fl. Germ.

PRENANTHES, Vaill.

P. alba, L. (Nabalus albus, Hook., Man.) WHITE LETTUCE.
Common. Aug.-Sept.

P. altissima, L. (Nabalus altissimus, Hook., Man.) TALL WHITE LETTUCE.
Common. Aug.-Sept.

P. serpentaria, Pursh. (Nabalus Fraseri, DC., Man.) LION'S FOOT. GALL-OF-THE-EARTH.
Tewksbury and Winchester (Dr. C. W. Swan); Belmont and Arlington (C. E. Perkins); Medford (Wm. Boott); Waltham List; Concord (Minot Pratt). Not common. Aug.-Sept.

TARAXACUM, Hall.

T. officinale, Weber. (T. Dens-leonis, Desf., Man.) DANDELION.
Very common. Apr.-July.

LACTUCA, Tourn.

L. Canadensis, L. WILD LETTUCE.
Common. Aug.-Sept.

L. integrifolia, Bigel. (L. Canadensis, var. integrifolia, Torr. & Gray, Man.)
Tewksbury (Dr. C. W. Swan); Malden and Melrose (F. S. Collins). Not very common. Aug.-Sept.

L. hirsuta, Muhl. (L. Canadensis, var. sanguinea, Torr. & Gray, Man.)
Lowell (Dr. C. W. Swan); Malden, Melrose, et al., not rare (F. S. Collins). Aug.-Sept.

L. SCARIOLA, L. PRICKLY LETTUCE.
Malden (F. S. Collins); Somerville (C. E. Perkins); Cambridge (Wm. Boott). Not common. Aug. Nat. from Eu.

L. leucophæa, Gray. (Mulgedium leucophaeum, DC., Man.) BLUE LETTUCE.
Widely distributed; not uncommon. Aug.-Sept.

SONCHUS, L.

S. OLERACEUS, L. SOW-THISTLE.
Malden, Somerville, Westford, Framingham, et al. Not uncommon. Nat. from Eu.

S. ASPER, Vill. SPINY-LEAVED SOW-THISTLE.
Malden, Medford, Lowell, et al. Not common. Nat. from Eu.

LOBELIACEÆ. LOBELIA FAMILY.

LOBELIA, L.

L. cardinalis, L. CARDINAL FLOWER.
Frequent. A form with white flowers is found in Melrose (Mrs. P. D. Richards), and Bedford (C. W. Jenks). July-Aug.
*****L. syphilitica,** L. GREAT LOBELIA.
Concord, introduced from Ill. by Minot Pratt. July-Aug.
L. inflata, L. INDIAN TOBACCO.
Common. July-Sept.
L. spicata, Lam.
Generally distributed, but not very common. July-Sept.
L. Dortmanna, L. WATER LOBELIA.
Fresh Pond, Martin's Pond, Dracut, Groton, Westford, et al. Scarce. July-Aug.

CAMPANULACEÆ. CAMPANULA FAMILY.

CAMPANULA, TOURN.

C. rotundifolia, L. HAREBELL.
Rarely found native, except in the Merrimac River Valley, where it is common. At Concord, it was introduced by Minot Pratt thirty years ago, and has become thoroughly established. June-Sept.
C. aparinoides, Pursh. MARSH BELL-FLOWER.
Common. June-Aug.
C. rapunculoides, L.
Tewksbury (Dr. C. W. Swan); Melrose and Natick (Rev. Thos. Morong); Malden (F. S. Collins). Adv. from Eu. July-Aug.

SPECULARIA, Heist.

S. perfoliata, A.DC. VENUS'S LOOKING-GLASS.
Rather common. June-July.

ERICACEÆ. HEATH FAMILY,

GAYLUSSACIA, HBK.

G. dumosa, Torr. & Gray. DWARF HUCKLEBERRY.
Melrose (H. A. Young); Natick (Rev. Thos. Morong); Littleton (L. L. Dame); Tewksbury and Newton (Wm. Boott). June.

G. frondosa, Torr. & Gray. BLUE TANGLE. DANGLEBERRY.
Widely distributed, but not very common. May-June.

G. resinosa, Torr. & Gray. BLACK HUCKLEBERRY.
Common. May-June.

VACCINIUM, L.

V. Oxycoccus, L. SMALL CRANBERRY.
Ashby and Wilmington (W. H. Manning); Concord (Minot Pratt); Natick (Rev. Thos. M rong); Littleton (L. L. Dame). Rare. June.

V. macrocarpon, Ait. CRANBERRY.
Common. June.

*** V.** *Vitis-Idœa,* L. COWBERRY.
Concord, introduced from the White Mts. by Minot Pratt, but does not thrive. June.

V. Pennsylvanicum, Lam. DWARF BLUEBERRY.
Common. A form with b'ack and shining berries destitute of bloom, the var. **nigrum** of Wood, occasional. May.

*** V. Canadense,** Kalm. CANADA BLUEBERRY.
Townsend (Miss H. E. Haynes). May.

V. vacillans, Soland. LOW BLUEBERRY.
Rather common. May-June.

V. corymbosum, L. SWAMP BLUEBERRY.
Common. May-June.

V. corymbosum, var. **atrococcum,** Gray.
Common. May-June.

CHIOGENES, Salisb.

C. hispidula, Torr. & Gray. CREEPING SNOWBERRY.
Shirley and Littleton (W. H. Manning); Tewksbury (B. D. Greene); Concord (Minot Pratt). Rare. May.

ARCTOSTAPHYLOS, Adans.

A. Uva-ursi, Spreng. BEARBERRY.
Widely distributed; but not very common. May-June.

EPIGÆA, L.

E. repens, L. TRAILING ARBUTUS. MAY FLOWER. Ashby, Littleton, Westford. Concord, et al. Common only in the north-western part of the county. Apr.-May.

GAULTHERIA, Kalm.

G. procumbens, L. WINTERGREEN. CHECKERBERRY. BOX-BERRY. Common. July.

LEUCOTHOE, Don.

L. racemosa, Gray.
Malden, Medford, Townsend, Ashland, et al. Not common. June.

CASSANDRA, Don.

C. calyculata, Don. LEATHER-LEAF.
Common. April-May.

ANDROMEDA, L.

A. polifolia, L.
Tewksbury, Dracut, Reading, Concord, Littleton, et al. Not common. June.

A. ligustrina, Muhl.
Common. June-July.

CLETHRA, L.

C. alnifolia, L. WHITE ALDER. SWEET PEPPERBUSH.
Common. July-Aug.

CALLUNA, Salisb.

C. vulgaris, Salisb. HEATHER.
This rare plant was discovered for the first time upon the American mainland, by Jackson Dawson, in July, 1861, on low grounds in Tewksbury. A second station has since been found in W. Andover, Essex Co., five miles from the Tewksbury locality, of the "green and smoothish variety precisely like the Tewksbury plant," Gray. Subsequent discoveries of the Calluna at many different points northward favor the conclusion that it is really indigenous in America. Mr. Dawson states that the Tewksbury station, never a large one, is decreasing in area, partly from the inroads of botanists, and partly from the encroachments of the woods beneath the shade of which the heather almost always disappears. The forbearance of collectors with the timely removal of a few shrubs would undoubtedly preserve this interesting survival of a former flora. July-Aug.

KALMIA, L.

K. latifolia, L. MOUNTAIN LAUREL.
Common in the northern towns, but rare elsewhere. June.

K. angustifolia, L. SHEEP LAUREL. LAMBKILL.
Common. June.

K. glauca, Ait. PALE LAUREL.
Ashby and Littleton (L. L. Dame); Westford (Misses Fletcher and Hodgman); Concord (G. E. Davenport); Natick (Austin Bacon); Groton and Carlisle (C. W. Jenks); Acton (Walter Deane). May.

RHODODENDRON, L.

R. viscosum, Torr. (Azalea viscosa, L., Man.) WHITE SWAMP HONEYSUCKLE. SWAMP PINK.
Common. June-July.

R. viscosum, Torr., var. **glaucum,** Gray.
Concord (H. S. Richardson); Tewksbury (L. L. Dame); Medford (G. E. Davenport). June-July.

R. nudiflorum, Torr. (Azalea nudiflora, L., Man.) PINK AZALEA. PINXTER-FLOWER.
Concord, Ashland, Framingham, Townsend, et al. Rare in the eastern part of the county. May-June.

R. Rhodora, Don. (Rhodora Candadensis, L., Man.)
Frequent. May.

LEDUM, L.

L. latifolium, Ait. LABRADOR TEA.
Townsend (E. H. Hitchings); Concord (G. E. Davenport); Natick (Austin Bacon); Littleton (W. H. Manning). Rare. May-June.

PYROLA, Tourn.

P. rotundifolia, L.
Common. June-July.

*****P. rotundifolia,** L. var **asarifolia,** Hook.
Concord (Thoreau); Stoneham (Wm. Boott). July.

P. elliptica, Nutt. SHIN-LEAF.
Common. June-July.

P. chlorantha, Swartz.
Common. June-July.

P. secunda, L.
Rather common. July.

MONESES, Salisb.

M. uniflora, Gray. ONE-FLOWERED PYROLA.
Stoneham (C. E. Dotey); Reading (C. E. Perkins); Natick (Austin Bacon); Townsend (Miss H. E. Haynes). Scarce. June.

CHIMAPHILA, Pursh.

C. umbellata, Nutt. PRINCE'S PINE. PIPSISSEWA.
Common. July.

C. maculata, Pursh. SPOTTED WINTERGREEN.
Generally distributed, but not common. Seldom more than a few plants in one locality. July.

MONOTROPA, L.

M. uniflora, L. INDIAN PIPE.
Common. June-July.

M. Hypopitys, L. PINE-SAP.
Common. June-July.

ILICINEÆ. (AQUIFOLIACEÆ, MAN.)

ILEX, L.

I. **verticillata,** Gray. BLACK ALDER. WINTERBERRY.
Common. June.

I. **lævigata,** Gray. SMOOTH WINTERBERRY.
Arlington, Cambridge and Lexington (Wm. Boott); Concord (Minot Pratt). Scarce. June.

NEMOPANTHES, Raf.

N. **Canadensis,** DC. MOUNTAIN HOLLY.
Rather common. May.

PLANTAGINACEÆ. PLANTAIN FAMILY.

PLANTAGO, L.

P. MAJOR, L. PLANTAIN.
Very common. June-Sept. Nat. from Eu. A form with panicled inflorescence found in Cambridge by C. E. Richardson is represented in the county herbarium. A form with very large thin leaves, lower side scabrous, found in Medford, 1866, by Wm. Boott, is in the Gray Herb. Nat. from Eu.

P. **Rugelii,** Decne. (P. Kamtschatica of the Man.)
June-Sept. Very common, and usually confounded with the last species, from which it is distinguished by commonly thinner leaves, with often reddish petioles, longer and more tapering spikes, narrower oblong sepals, capsules opening below the middle, and seeds without reticulation.

P. **decipiens,** Barn. (P. maritima, L., var. juncoides, Gray. Man.) SEASHORE PLANTAIN.
Common on salt marshes.

P. LANCEOLATA, L. RIBGRASS. ENGLISH PLANTAIN.
Common. June-July. Nat. from Eu.

P. *Patagonica,* Jacq., var. *spinulosa,* Gray.
Somerville (C. E. Perkins). Adv. from farther West.
"A canescent form with aristately prolonged and rigid bracts."
Syn. Fl. N. A.

P. *Patagonica,* Jacq., var. *aristata,* Gray.
Malden (F. S. Collins). Adv. from farther West.

PLUMBAGINACEÆ. LEADWORT FAMILY.

STATICE, Tourn.

S. Limonium, L., var. **Caroliniana,** Gray. SEA LAVENDER.
MARSH ROSEMARY.
Very common on salt marshes. Aug.-Sept.

PRIMULACEÆ. PRIMROSE FAMILY.

DODECATHEON, L.

*D. *Meadia*, L. AMERICAN COWSLIP. SHOOTING STAR.
Concord, introduced from the West by Minot Pratt. May-June.

TRIENTALIS, L.

T. Americana, Pursh. STAR-FLOWER. CHICKWEED WINTER-GREEN.
Common. May-June.

LYSIMACHIA, Tourn.

L. thyrsiflora, L. TUFTED LOOSESTRIFE.
Rather common, at least in the eastern part of the county. June-July.

L. stricta, Ait.
Common. July-Aug.

L. quadrifolia, L.
Common. June-July.

L. NUMMULARIA, L. MONEYWORT.
Medford (C. E. Perkins); Stoneham (F. S. Collins); Townsend (Miss H. E. Haynes). July-Aug. Introduced from Eu., and sparingly naturalized.

L. VULGARIS, L.
Stoneham, roadside (F. S. Collins); abundantly naturalized in a swamp near Spy Pond (L. H. Bailey, Jr.) Specimen in Gray Herb. Aug. For description, see Gray, School and Field Book.

STEIRONEMA, Raf.

S. ciliatum, Raf. (Lysimachia ciliata, L., Man.)
Found throughout the county, but rare in the eastern part. July.

S. lanceolatum, Gray. (Lysimachia lanceolata, Walt., Man).
Common. June-July.

*S. lanceolatum, Gray, var. hybridum, Gray.
Concord, common (Minot Pratt). June-July.

S. lanceolatum, Gray, var. **angustifolium,** Gray.
Malden (R. Frohock); Concord, common (Minot Pratt); Newton
(F. S. Collins). June-July.

GLAUX, L.

G. maritima, L. SEA-MILKWORT.
Somerville, bank of the Mystic (F. S. Collins). Scarce. June.

ANAGALLIS, Tourn.

A. arvensis, L. PIMPERNEL.
Medford (Wm. Boott); Malden (H. A. Young). Rare. June-Sept.
Nat. from Eu.

SAMOLUS, L.

S. Valerandi, L., var. **Americanus,** Gray. WATER PIMPERNEL. BROOKWEED.
Medford (Wm. Boott); Winchester (C. E. Perkins). Rare.
June-Aug.

HOTTONIA, L.

H. inflata, Ell. FEATHERFOIL.
Malden (F. S. Collins); Medford (G. E. Davenport); Stoneham
(L. L. Dame). June-July. Not reported outside the limits of
Middlesex Fells, save at Hammond's Pond, by Wm. Boott.

LENTIBULARIACEÆ. BLADDERWORT FAMILY.

UTRICULARIA, L.

U. inflata, Walt. INFLATED BLADDERWORT.
Rather common. July-Aug.

U. vulgaris, L., var. **Americana,** Gray. GREATER BLADDERWORT.
Rather common. July-Aug.

U. minor, L. SMALLER BLADDERWORT.
Tewksbury (Wm. Boott); Spot Pond, Medford (Rev. Thomas
Morong); Reading (*fide* specimen in herb, B. S. N. H.) June-July.

U. intermedia, Hayne.
Reading, Cambridge, Ashland, Townsend, et al. Not uncommon.
July-Aug.

U. gibba, L.
Spot Pond, Medford (G. E. Davenport); Hammond's Pond, Newton (C. E. Faxon); Waushakum Pond, Ashland (Rev. Thos.
Morong); Flat Pond, Groton (Dr. C. W. Swan); Westford (W. H.
Manning). Widely distributed, but infrequent. July-Sept.

U. purpurea, Walt. PURPLE BLADDERWORT.
Generally distributed, but not common. Aug.-Sept.

U. resupinata, Greene.
Round Pond, Tewksbury, Aug. 24. 1865; Silver Lake, Wilmington, Aug. 4, 1869 (Wm. Boott). Very rare.

U. cornuta, Michx.
Common. July-Aug.

OROBANCHACEÆ. BROOM-RAPE FAMILY.

EPIPHEGUS, Nutt.

E. Virginiana, Bart. BEECH-DROPS.
Arlington, Cambridge and Winchester (Wm. Boott); Waltham List; Framingham (Rev. J. H. Temple). Not common. Aug.-Sept.

APHYLLON, Mitch.

A. uniflorum, Gray. ONE-FLOWERED CANCER-ROOT. BROOM-RAPE.
Common. June.

SCROPHULARIACEÆ. FIGWORT FAMILY.

VERBASCUM, L.

V. THAPSUS, L. MULLEIN.
Common. July-Sept. Nat. from Eu.

V. BLATTARIA, L. MOTH MULLEIN.
Malden (F. S. Collins); Concord (Minot Pratt); Stoneham (Mrs. P. D. Richards). Scarce. June-July. Nat. from Eu.

V. Lychnitis, L. WHITE MULLEIN.
Westford (W. E. Coburn). June-Aug. Adv. from Eu.

LINARIA, Tourn.

L. Canadensis, Dumont. TOAD-FLAX.
Common. A specimen from Chelmsford, collected by Dr. F. Nickerson, has the regular five-parted flower of the Peloria state. June-Sept.

L. VULGARIS, Mill. TOAD-FLAX. BUTTER-AND-EGGS.
Common. July-Sept. Nat. from Eu.

L. Cymbalaria, Mill. COLISEUM-IVY.
Lowell, "dumps" (Dr. C. W. Swan); Malden (F. S. Collins). Adv. from Eu. For description, see Wood's Bot. & Fl.

SCROPHULARIA, Tourn.

S. nodosa, L. FIGWORT.
Chelmsford (Dr. C. W. Swan); Stoneham (C. E. Dotey); Watertown (C. E. Perkins); Cambridge (Walter Deane); Concord (Minot Pratt). Rare. June-July.

CHELONE, Tourn.

C. glabra, L. SNAKE-HEAD.
Common. July-Sept.

PENTSTEMON, Mitch.

P. pubescens, Soland.
Framingham (Rev. J. H. Temple); Townsend, 1886 (E. H. Hitchings). Rare. July-Aug.

MIMULUS, L.

M. ringens, L. MONKEY-FLOWER.
Common. July-Sept.

M. brevipes, Benth.
Wool-waste, N. Chelmsford (Rev. W. P. Alcott; specimen in herb. of). Adv. from Cal. For description, see Gray, Syn. Fl. N. A.

GRATIOLA, L.

G. Virginiana, L.
Mystic Pond, Medford (Wm. Boott); Somerville (C. E. Perkins); Concord (Minot Pratt). Scarce. July-Sept.

G. aurea, Muhl. HEDGE-HYSSOP.
Common. The white variety has been found at Winchester, by W. H. Manning; both the white and light yellow varieties at Westford, by Dr. C. W. Swan. July-Sept.

ILYSANTHES, Raf.

I. gratioloides, Benth. FALSE PIMPERNEL.
Rather common. July-Aug.

VERONICA, L.

V. Americana, Schwein. AMERICAN BROOKLIME.
Westford (Dr. C. W. Swan). Rare. June-Aug.

V. scutellata, L. MARSH SPEEDWELL.
Rather common. June-Sept.

V. officinalis, L. SPEEDWELL.
Westford (Miss Emily F. Fletcher); Arlington (L. L. Dame); Natick (Austin Bacon); Concord (Minot Pratt); Acton (Walter Deane). Scarce. June-July.

V. serpyllifolia, L. THYME-LEAVED SPEEDWELL.
Common. May-Sept.

V. peregrina, L. NECKWEED. PURSLANE SPEEDWELL.
Somerville, Cambridge, Belmont, Concord, et al. Rather common.
May-June.
V. ARVENSIS, L. CORN SPEEDWELL.
Rather common. May-June. Nat. from Eu.
V. spicata, L.
Roadside, N. Chelmsford (Dr. C. W. Swan). Adv. from Eu. For description, see Wood's Bot. & Fl.
V. AGRESTIS, L. FIELD SPEEDWELL.
Reading (C. E. Perkins); Framingham (Rev. J. H. Temple). Rare. Nat. from Eu.

GERARDIA, L.

G. purpurea, L., var. **paupercula,** Gray. PURPLE GERARDIA.
Common. A form with white flowers has been reported at Lowell (Dr. C. W. Swan); Malden (F. S. Collins); and Concord (Minot Pratt). July-Sept.
G. maritima, Raf. SEASIDE GERARDIA.
Common on salt marshes; Concord, introduced by Minot Pratt. Aug.
G. tenuifolia, Vahl. SLENDER GERARDIA.
Common. A form with white flowers is found at Malden (F. S. Collins). July-Sept.
G. flava, L. DOWNY FALSE FOXGLOVE.
Rather common. July-Aug.
G. quercifolia, Pursh. SMOOTH FALSE FOXGLOVE.
Generally distributed, but not very common. July-Aug.
G. pedicularia, L.
Common. Aug-Sept.

CASTILLEIA, Mutis.

C. coccinea, Spreng. PAINTED CUP.
Throughout the county, but not common in the towns near Boston. A variety with yellow bracts in Tewksbury and Dracut (Rev. J. L. Russell, Hovey's Mag., Vol. VII). May-June.

ORTHOCARPUS, Nutt.

O. purpurascens, Benth.
Wool-waste, N. Chelmsford (Rev. W. P. Alcott: specimen in herb. of). Adv. from coast of Cal. For description, see Gray, Syn. Fl. N. A.

PEDICULARIS, Tourn.

P. Canadensis, L. LOUSEWORT. WOOD BETONY.
Common. May-June.

MELAMPYRUM, Tourn.

M. Americanum, Michx. COW-WHEAT.
Common. May-Sept.

VERBENACEÆ. VERVAIN FAMILY.

VERBENA, L.

V. hastata, L. BLUE VERVAIN.
Common. A form with pink flowers at Winchester (Mrs. P. D. Richards). June-Sept.

V. urticæfolia, L. WHITE VERVAIN.
Common. July-Sept.

*V. officinalis, L. EUROPEAN VERVAIN.
Framingham (Rev. J. H. Temple). Adv. from Eu.

V. bracteosa, Michx.
Lowell (Dr. C. W. Swan); Cambridge (C. E. Perkins). Aug. Adv. from the West.

PHRYMA, L.

P. Leptostachya, L. LOPSEED.
Malden, Arlington, Woburn, et al. Not common. June-July.

LABIATÆ. MINT FAMILY.

TEUCRIUM, L.

T. Canadense, L. GERMANDER. WOOD SAGE.
Lowell (Dr. C. W. Swan); Framingham (Rev. J. H. Temple); Cambridge, Watertown, Medford, et al. Not uncommon near salt water, but rare elsewhere. July-Aug.

TRICHOSTEMA, L.

T. dichotomum, L. BLUE CURLS.
Common. July-Sept.

MENTHA, L.

M. VIRIDIS, L. SPEARMINT.
Rather common. July-Aug. Nat. from Eu.

M. PIPERITA, L. PEPPERMINT.
Rather common. July-Sept. Nat. from Eu.

M. sativa, L. WHORLED MINT.
Chelmsford, roadside (Dr. C. W. Swan). Adv. from Eu.

M. ARVENSIS, L. CORN MINT.
Cambridge, (fide specimen in Gray Herb.); Wakefield (F. S. Collins); Nat. from Eu.

M. Canadensis, L. WILD MINT.
Common. July-Sept.

LYCOPUS, L.

L. Virginicus, L. BUGLE-WEED.
Rather common. Aug.-Sept.

L. sinuatus, Ell. (L. Europæus, L., var. sinuatus, Man.)
Common. July-Sept.

PYCNANTHEMUM, Michx.

P. incanum, Michx.
Lowell (Dr. C. W. Swan); Medford (C. E. Perkins); Winchester (Mrs. P. D. Richards); Framingham (Rev. J. H. Temple); Concord (Minot Pratt). Not very common. July-Aug.

P. clinopodioides, Gray.
Concord (Walter Deane; specimen in herb. of). Aug.-Sept. Adv. from farther south.

P. muticum, Pers.
Widely distributed, but not very common. Aug.

P. lanceolatum, Pursh.
Melrose (R. Frohock); Winchester (Mrs. P. D. Richards); Concord (Minot Pratt). Scarce. Aug.-Sept.

P. linifolium, Pursh.
Reading (F. H. Gilson). An unusually northern station for this species. Aug.-Sept.

THYMUS, L.

*T. serpyllum, L. CREEPING THYME.
Malden (C. E. Perkins). Rare. Aug. Adv. from Eu.

SATUREIA, L.

S. hortensis, L. SUMMER SAVORY.
" Dumps," occasional. Adv. from Eu.

CALAMINTHA, Moench.

*C. Clinopodium, Benth. BASIL.
Framingham (Rev. J. H. Temple); Concord (Minot Pratt). July. Adv. from Eu.

HEDEOMA, Pers.

H. pulegioides, Pers. PENNYROYAL.
Common. July-Sept.

H. hispida, Pursh.
Reading (W. H. Manning). July. Adv. from the West.

COLLINSONIA, L.

C. Canadensis, L. RICH-WEED. STONE-ROOT.
Ashby (L. L. Dame); Dracut (Dr. C. W. Swan). Rare. July-Aug.

SALVIA, L.

S. tiliæfolia, Vahl.
Cambridge, rubbish-heap, 1885 (Walter Deane; specimen in herb. of). Adv. from S. America.
"Stem herbaceous, erect, smoothish or slightly pubescent, leaves broadly ovate, crenate, truncate or sub-cordate at the base, softly wrinkled and pubescent, with scattered hairs; the floral leaves membranaceous, lanceolate, deciduous; racemes simple; whorls loose, approximate, 6-10 flowered; calyx tubular with ciliate nerves; teeth 3, ovate-lanceolate, acute; corolla scarcely exceeding the calyx; lobes of the style subequal, subulate, or the upper longer; corolla blue." DC. Prodr. XII, p. 299.

MONARDA, L.

*M. didyma, L. OSWEGO TEA.
Concord, escaped (Minot Pratt); Framingham (Rev. J. H. Temple). Rare. July. Adv. from farther West.

M. fistulosa, L. BERGAMOT.
Lowell (Dr. C. W. Swan); Medford and Winchester (Mrs. P. D. Richards); Melrose (Rev. Thos. Morong); Concord, introduced from Ill. by Minot Pratt. Rare. July-Aug. A form with crimson flowers, approaching var. **rubra** has been reported at Littleton, a possible hybrid.

M. fistulosa, L., var. **mollis,** Benth.
Lowell (Dr. C. W. Swan). July-Aug.
"Corolla from flesh-color to lilac, glandular, and its upper lip hairy outside, or more bearded at the tip: leaves paler, soft-pubescent beneath, often shorter petioled." Syn. Fl. N. A.

BLEPHILIA, Raf.

B. ciliata, Raf.
Ashland (Rev. Thos. Morong; specimen in herb. of). Very rare in New England, common farther south. June-July.

LOPHANTHUS, Benth.

*L. anisatus, Benth.
Concord, introduced from Wisconsin by Minot Pratt.

NEPETA, L.

N. Cataria, L. CATNIP.
Common. Roadsides, and occasionally scattered over adjacent pastures. July-Aug. Nat. from Eu.

N. GLECHOMA, Benth. GROUND IVY.
Common. May-July. Nat. from Eu.

PHYSOSTEGIA, Benth.

P. *Virginiana*, Benth. FALSE DRAGON-HEAD.
Stoneham (Mrs. A. M. Moody); also growing in a thicket in the same town (L. L. Dame). Probably an escape from cultivation. Aug.-Sept.

BRUNELLA, Tourn.

B. **vulgaris**, L. SELF-HEAL. HEAL-ALL.
Very common. A white variety at Westford (Dr. C. W. Swan). July-Sept.

SCUTELLARIA, L.

S. **galericulata**, L. SKULLCAP.
Rather common. July-Aug.

S. **lateriflora**, L. MAD-DOG SKULLCAP.
Rather common. July-Aug.

MARRUBIUM, L.

M. VULGARE, L. HOREHOUND.
Occasionally spontaneous. June-July. Nat. from Eu.

GALEOPSIS, L.

G. TETRAHIT, L. HEMP-NETTLE.
Lowell and Tewksbury (Dr. C. W. Swan); Natick (Austin Bacon); Malden, station now destroyed, and Woburn (F. S. Collins). July-Aug. Nat. from Eu.

STACHYS, L.

S. **aspera**, Michx. (S. palustris, L., var. aspera of the Man.)
Concord, Lowell and Bedford (Dr. C. W. Swan); Framingham (Miss J. W. Williams); Medford (Wm. Boott). July-Sept.

S. *Betonica*, Benth. (Betonica officinalis, L., Man.) WOOD BETONY.
"Found by C. J. Sprague in a thicket at Newton, Mass." (Man.) Specimen in herb. B. S. N. H. Adv. from Eu.

LEONURUS, L.

L. CARDIACA, L. MOTHERWORT.
Common. July-Sept. Nat. from Eu.

LAMIUM, L.

L. AMPLEXICAULE, L. DEAD-NETTLE.
Medford and Stoneham (L. L. Dame); Waltham List; Concord (Minot Pratt). Rare. Apr.-Sept. Nat. from Eu.

L. ALBUM, L. WHITE DEAD-NETTLE.
Somerville and Cambridge (C. E. Perkins); et al. Rare. Nat. from Eu.

L. *maculatum*, L.
Medford, escaped from cultivation. Apr.-Sept. Int. from Eu. For description, see Wood's Bot. & Fl.

BALLOTA, L.

B. *nigra*, L. BLACK HOREHOUND.
Cambridge, 1885 (Walter Deane; specimen in herb. of). July.

BORRAGINACEÆ. BORAGE FAMILY.

BORRAGO, Tourn.

B. *officinalis*, L. BORAGE.
Lowell, 1848 (J. A. Lowell), specimen in herb. B. S. N. H. July-Sept. Adv. from Eu. For description, see Wood's Bot. & Fl.

ECHIUM, Tourn.

E. VULGARE, L. VIPER'S BUGLOSS.
Medford (Wm. Boott); Somerville (C. E. Perkins); Cambridge (Walter Deane); Ashland (Rev. Thos. Morong). June. Nat. from Eu.

LYCOPSIS, L.

L. *arvensis*, L. SMALL BUGLOSS.
Lowell, wayside (Dr. F. Nickerson). Adv. from Eu.

SYMPHYTUM, Tourn.

S. OFFICINALE, L. COMFREY.
Stoneham, escaped (L. L. Dame); Newton (C. E. Perkins); Concord (Minot Pratt); Groton (C. W. Jenks). June-July. Nat. from Eu.

S. ASPERRIMUM, Sims.
Ashland, escaped and sparingly established (Rev. Thos. Morong; specimen in herb. of). Nat. from Eu.
"Stems branching, rough with stiff, somewhat recurved aculeate prickles; leaves ovate-lanceolate, very pointed at both ends, rough, the lower petioled, the upper subsessile; calyx divided above the middle into rough, subulate lobes; corolla campanulate, four times the length of the calyx; lanceolate appendages of the length of the stamens, ciliate-papillose; style included; flowers, bluish purple."
DC. Prodr.

LITHOSPERMUM, Tourn.

L. ARVENSE, L. CORN GROMWELL.
Lowell, a garden weed (Dr. C. W. Swan); Somerville and Medford (C. E. Perkins); Reading (W. H. Manning). Not common. June-Aug. Nat. from Eu.

*L. angustifolium, Michx. (L. longiflorum, Spreng., Man.)
Concord, introduced from Ill. by Minot Pratt. May.

MERTENSIA, Roth.

M. Virginica, DC. VIRGINIAN COWSLIP. LUNGWORT.
Concord, introduced from Ill. by Minot Pratt. June.

MYOSOTIS, L.

M. PALUSTRIS, With. TRUE FORGET-ME-NOT.
Woburn, abundant in a brook near new line of B. & L. R. R., and also completely filling a section of the bed of the old Middlesex Canal: thoroughly established (L. L. Dame). June-Aug. Nat. from Eu.

M. laxa, Lehm. (M. palustris, With., var. laxa, Man.)
Rather common. June-Aug.

***M. arvensis**, Hoffm.
Chelmsford (Bigelow's Fl. Bost.); Malden (F. S. Collins); Marlboro (Mrs. A. M. Staples); Groton (C. W. Jenks). Scarce. June-Aug.

M. verna, Nutt.
Common. June-July.

AMSINCKIA, Lehm.

A. intermedia, Fisch. & Meyer.
Wool-waste, N. Chelmsford (Rev. W. P. Alcott); Lowell, wool-waste (Dr. F. Nickerson). Adv. from the Pacific coast. For description, see Gray, Syn. Fl. N. A.

ERITRICHIUM, Schrad.

E. oxycaryum, Gray.
Wool-waste, N. Chelmsford (Rev. W. P. Alcott). For description, see Gray, Syn. Fl. N. A.

ECHINOSPERMUM, Swartz.

E. Lappula, Lehm. STICKSEED.
Westford, woollen mills (Dr. C. W. Swan); Malden, old shoddy mill, station now destroyed (F. S. Collins); Cambridge (Walter Deane). Rare. July-Aug. Adv. from Eu.

E. Virginicum, Lehm. (Cynoglossum Morisoni, DC., Man.) BEGGAR'S LICE.
Westford and Lowell (Dr. C. W. Swan); Arlington (Wm. Boott); Weston and Stoneham (F. S. Collins); Concord (Minot Pratt). Not common. June-July.

CYNOGLOSSUM, Tourn.

C. OFFICINALE, L. HOUND'S TONGUE.
Lowell, woollen mills (Dr. C. W. Swan); Malden (F. S. Collins); Stoneham (L. L. Dame); Belmont (H. S. Richardson); Concord (Minot Pratt). Not common. June-July. Nat. from Eu.

HELIOTROPIUM, Tourn.

H. Europæum, L. HELIOTROPE.
Cambridge, 1884 (Walter Deane); Westford, woollen mill yard (Dr. F. Nickerson).

H. Indicum, L. (Heliophytum Indicum, DC., Man.) INDIAN HELIOTROPE.
Cambridge, Sept., 1884 (Walter Deane; specimen in herb. of). Adv. from India.

ASPERUGO, L.

A. procumbens, L.
Malden and Somerville. "dumps" (F. S. Collins). Adv. from Eu., tending towards naturalization.
"Nutlets 4, compressed, adnate at the side to the narrow style; fruiting calyx compressed; sinuses plane, parallel, sinuate." Koch, Syn. Fl. Germ.

HYDROPHYLLACEÆ. WATERLEAF FAMILY.

HYDROPHYLLUM, L.

*H. Virginicum, L. WATERLEAF.
Concord, introduced from Vermont by Minot Pratt. June.

PHACELIA, Juss.

*P. congesta, Hook.
Cambridge, "dump" (Rev. Thos. Morong). June. Adv. from Texas. For description, see Wood's Bot. & Fl.

P. brachyloba, Gray.
N. Chelmsford, wool-waste (Rev. W. P. Alcott; specimen in herb. of). June. Adv. from Cal. For description, see Gray, Syn. Fl. N. A.

P. Whitlavia, Gray.
N. Chelmsford, wool-waste, abundant (Rev. W. P. Alcott; specimen in herb. of). Adv. from Cal. For description, see Gray, Syn. Fl. N. A.

P. tanacetifolia, Benth.
N. Chelmsford, wool-waste (Rev. W. P. Alcott; specimen in herb. of). Adv. from Cal. For description, see Wood's Bot. & Fl.

P. circinata, Jacq. f.
N. Chelmsford, wool-waste (Rev. W. P. Alcott; specimen in herb. of). Adv. from Cal. For description, see Gray, Syn. Fl. N. A.

POLEMONIACEÆ. PHLOX FAMILY.

PHLOX, L.

P. paniculata, L.
Melrose, occasionally escaped (Rev. Thos. Morong). June-July, Native farther South.

GILIA, Ruiz. & Pav.

G. leucocephala, Gray.
N. Chelmsford, wool-waste (Rev. W. P. Alcott; specimen in herb. of). Adv. from Cal. For description, see Gray, Syn. Fl. N. A.

G. inconspicua, Dougl.
N. Chelmsford, wool-waste (Rev. W. P. Alcott; specimen in herb. of). Adv. from Cal. For description see Gray, Syn. Fl. N. A.

COVOLVULACEÆ. CONVOLVULUS FAMILY.

IPOMŒA, L.

I. purpurea, Lam. MORNING-GLORY.
Occasionally found escaped. July-Oct. Adv. from Trop. Am.

I. Nil, Roth.
Lowell, escaped (Dr. C. W. Swan). July-Oct. Native of Eu.

I. hederacea, Jacq.
Lowell, "dump" (Dr. C. W. Swan). Adv. from Trop. Am. For description see Wood's Bot. & Fl., under Pharbitis hederacea.

I. lacunosa, L.
Lowell, "dump" (Dr. C. W. Swan). Aug.-Sept. Adv. from the West.

I. commutata, Roem. & Sch.
Lowell, "dump" (Dr. C. W. Swan). July-Sept. Adv. from farther South. For description, see Wood's Bot. &. Fl.

CONVOLVULUS, L.

C. ARVENSIS, L. BINDWEED.
Malden (F. S. Collins); Medford (G. E. Davenport); Marlboro (Mrs. A. M. Staples); Concord (Minot Pratt). Not uncommon. June-July. Nat. from Eu.

C. sepium, L. (Calystegia sepium, R. Br., Man.) HEDGE BIND-
WEED.
Not uncommon. *Flore pleno,* escaped, in fields; Westford, Bedford and Lowell (Dr. C. W. Swan). June-Aug.

CUSCUTA, Tourn.

C. arvensis, Beyrich.
Winter Pond, Winchester (Dr. C. W. Swan). Rare. An extreme northern locality; host plant, Crotalaria sagittalis. June-July.

C. Gronovii, Willd. DODDER.
Common. Aug.-Sept.

SOLANACEÆ. NIGHTSHADE FAMILY.

LYCOPERSICUM, Tourn.

L. esculentum, Mill. TOMATO.
Waste places. June-Sept.

SOLANUM, Tourn.

S. tuberosum, L. POTATO.
Waste places. July-Aug. Native of S. Am.

S. DULCAMARA, L. BITTERSWEET. WOODY NIGHTSHADE.
Common. June-Aug. Nat. from Eu.

S. nigrum, L. NIGHTSHADE.
Lowell and Chelmsford (Dr. C. W. Swan); Medford (C. E. Perkins); Framingham (Rev. J. H. Temple); Concord (Minot Pratt). Rare. July-Sept. According to Gray, Syn. Fl. N. A., a cosmopolite. "common in damp or shady, especially cultivated or waste grounds, appearing as if introduced."

S. Carolinense, L. HORSE-NETTLE.
Watertown and Reading (C. E. Perkins). Rare. June-Aug.

S. sisymbriifolium, Lam.
Cambridge, rubbish heap, 1884 and 1885 (Walter Deane; specimen in herb. of). Adv. from S. Am.
"Stem somewhat herbaceous, hairy, viscid, prickly; leaves viscid, hairy, prickly on both sides, pinnatifid, lobes acute, sinuate-dentate, racemes terminal and lateral, calyx 5-angled, inflated, prickly, covering the berry." DC. Prodr., XIII, 1, 326.

S. rostratum, Dun.
Lowell, rather common (Dr. F. Nickerson); Watertown and Somerville (C. E. Perkins); Malden (F. S. Collins). Aug.-Sept. Adv. from the West. For description, see Wood's Bot. and Fl.

PHYSALIS, L.

P. Philadelphica, Lam.
Cambridge, rubbish heap, 1884 (Walter Deane; specimen in herb. of). Adv. from farther south.

P. pubescens, L.
Lowell, "dump" (Dr. C. W. Swan); Bedford (Miss A. Browne). Rare. Possibly introduced.

P. Virginiana, Mill. (P. viscosa, L., Man.)
Cambridge, rubbish heap (Walter Deane); Lowell, near mills (Dr. C. W. Swan). Rare. July-Sept.

P. capsicifolia, Dun.
Cambridge, 1884 (Walter Deane; specimen in herb. of). Sept. Adv. from S. Am. & Asia.
"Stem rather stout at the base, greenish purple, smooth, three-sided; branches obliquely ascending, the upper quadrangular; leaves long-petioled, oblong, subentire, pointed at each end, unequal at the base, thin, smooth; peduncles filiform, smoothish, erect during flowering, afterward pendulous; calyx salver-shaped, smooth, 5 parted, lobes deltoid, acuminate; corolla sinuately 5-angled, angles acute; anthers oblong, bluish; fruiting calyx roundish, 5-angled, when ripe greenish-yellow, with purple lines and streaks." DC. Prodr. XIII. I. 449.

NICANDRA, Adans.

N. PHYSALOIDES, Gaertn. APPLE OF PERU.
Not very common. July-Aug. Nat. from Peru.

PETUNIA, Juss.

P. nyctaginiflora, Juss.
Malden, waste heap, and elsewhere. July-Oct. Adv. from S. Am. For description, see Wood's Bot. & Fl.

LYCIUM, L.

L. VULGARE, Dun. MATRIMONY VINE.
Escaped sparingly. June-Aug. Nat. from Eu.

HYOSCYAMUS, Tourn.

H. NIGER, L. BLACK HENBANE.
Somerville (F. S. Collins); has perpetuated itself for many years along the Andover Turnpike. June-July. Nat. from Eu.

DATURA, L.

D. STRAMONIUM, L. THORN-APPLE.
Not very common. July-Sept. Nat. from Asia.

D. Tatula, L. Purple Thorn-Apple.
More common than the preceding. July-Sept. Nat. from Trop. Am.

D. inermis, Jacq.
Cambridge, rubbish heap, 1884 & 1885 (Walter Deane). July-Sept. Adv. from Africa.
"Stem branching, hollow, terete, smooth; leaves long petioled, smooth on both sides, acute, incised into acute lobes; flowers on short, winged petioles; calyx 5-angled, smooth; corolla twice the length of the calyx, with roundish, cuspidate lobes; capsule ovate, obtuse, smooth, unarmed, always erect, four valved." DC. Prodr. XIII. I. 539.

D. meteloides, DC.
Cambridge, rubbish heap (Walter Deane; specimen in herb. of). July-Sept. Adv. from Mexico. For description, see Wood's Bot. & Fl.

NICOTIANA, L.

N. Bigelovii, Wats.
Lowell, waste ground (Dr. C. W. Swan); N. Chelmsford, wool-waste (Rev. W. P. Alcott). Adv. from Cal. For description, see Gray, Syn. Fl. N. A.

GENTIANACEÆ. GENTIAN FAMILY.

SABBATIA, Adans.

*****S. chloroides,** Pursh.
Concord, introduced from Weymouth, Mass., by Minot Pratt. A form with white flowers was introduced with the type. July-Sept.

GENTIANA, L.

G. crinita, Froel. FRINGED GENTIAN.
Not uncommon save in the vicinity of cities, where it is becoming rare. A form with pink flowers at So. Sudbury (Geo. H. Whitney); a form with white flowers occasional. Sept.

G. Andrewsii, Griseb. CLOSED GENTIAN.
In most parts of the county, but less common than the preceding species. Aug.-Sept.

BARTONIA, Muhl.

B. tenella, Muhl.
In most parts of the county, but not very common. Aug.-Sept.

MENYANTHES, Tourn.

M. trifoliata, L. BUCKBEAN.
Not uncommon. May.

LIMNANTHEMUM, Gmel.

L. lacunosum, Griseb. FLOATING HEART.
Throughout the county, but nowhere very common. July-Aug.

APOCYNACEÆ. DOGBANE FAMILY.

APOCYNUM, Tourn.

A. androsæmifolium, L. DOGBANE.
Common. June-July.
A. cannabinum, L. INDIAN HEMP.
Everett, Woburn, Townsend, et al.; not common. June-July.

ASCLEPIADACEÆ. MILKWEED FAMILY.

ASCLEPIAS, L.

A. Cornuti, Decne. MILKWEED.
Very common. July-Aug.
A. phytolaccoides, Pursh. POKE MILKWEED.
Generally distributed, but not common. June-July.
A. purpurascens, L. PURPLE MILKWEED.
Not very common. July-Aug.
A. quadrifolia, L. FOUR-LEAVED MILKWEED.
Not very common. June.
A. incarnata, L., var. **pulchra,** Gray.
Common. July-Aug. The type, though often reported, does not appear to be within the county limits.
A. obtusifolia, Michx.
Not common. A form with the leaves in whorls of three was found at Concord by W. H. Manning. July-Aug.
A. obtusifolia × phytolaccoides, (*fide* Asa Gray).
Several plants were found along the banks of the old Middlesex Canal in Wilmington, July, 1885.
A. tuberosa, L. BUTTERFLY-WEED.
This plant, frequently mentioned by the older botanists, has become rare, at least in the vicinity of Boston. July-Aug.
A. verticillata, L. WHORLED MILKWEED.
Malden, Woburn, Framingham, Natick, et al. Not common. July-Sept.

VINCETOXICUM, Moench.
V. NIGRUM, Moench.
Watertown (C. E. Perkins); Cambridge (Man.); Medford and Ashland (Rev. Thos. Morong). June. Nat. from Eu.

PERIPLOCA, L.
*P. Græca, L.
Concord, introduced by Minot Pratt. Aug. A native of Eu.

OLEACEÆ. OLIVE FAMILY.

LIGUSTRUM, Tourn.
L. VULGARE, L. PRIVET.
Common. June. Nat. from Eu.

FRAXINUS, Tourn.
F. Americana, L. WHITE ASH.
Common. Apr.-May.
F. pubescens, Lam. RED ASH.
Occasional throughout the county. Apr.-May.
F. sambucifolia, Lam. BLACK or WATER ASH.
Medford, Weston, Framingham and Ashby (L. L. Dame); Concord (Walter Deane); Tewksbury (Dr. C. W. Swan). Not very common. Apr.-May.

SYRINGA, L.
S. VULGARIS, L. LILAC.
Along roadsides, near old houses. Often well established, of lower growth and smaller leaves than the cultivated plant. May-June. A native of Eu. For description, see Wood's Bot. and Fl.

ARISTOLOCHIACEÆ. BIRTHWORT FAMILY.

ASARUM, Tourn.
A. CANADENSE, L. WILD GINGER.
Westford, locally established (Miss E. F. Fletcher); the Concord plant was introduced from Vt. by Minot Pratt. May-June.

NYCTAGINACEÆ. FOUR-O'CLOCK FAMILY.

OXYBAPHUS, Vahl.
O. nyctagineus, Sweet.
Cambridge (C. E. Perkins). June-Aug. Adv. from the West.

PHYTOLACCACEÆ. POKEWEED FAMILY.

PHYTOLACCA, Tourn.

P. decandra, L. POKE. GARGET.
Common. July-Sept.

CHENOPODIACEÆ. GOOSEFOOT FAMILY.

CHENOPODIUM, L.

*C. polyspermum, L.
Framingham (Rev. J. H. Temple). July-Sept. Adv. from Eu.

C. ALBUM, L. PIGWEED.
Very common. July-Sept. Nat. from Eu.

C. URBICUM, L.
Somerville (C. E. Perkins); Lowell (Dr. C. W. Swan); Medford (F. S. Collins). July-Sept. Nat. from Eu.

C. URBICUM, L., var. RHOMBIFOLIUM, Moq.
Newton (C. J. Sprague); specimen in herb. B. S. N. H. July-Sept. Nat. from Eu.

C. MURALE, L.
Tewksbury (B. D. Greene), specimen in herb. B. S. N. H.; Framingham (Rev. J. H. Temple). Not common. July-Sept. Nat. from Eu.

C. HYBRIDUM, L. MAPLE-LEAVED GOOSEFOOT.
Common. July-Sept. Nat. from Eu.

C. BOTRYS, L. JERUSALEM OAK.
Lowell, Cambridge, Somerville, et al. Introduced from Eu., and apparently naturalized near Boston, where it has been not uncommon since Bigelow's Fl. Bost. July-Sept.

C. AMBROSIOIDES, L. MEXICAN TEA.
Somerville (C. E. Perkins); Cambridge and Medford (F. S. Collins). July-Sept. Int. from Trop. Am.

C. AMBROSIOIDES, L., var. ANTHELMINTICUM, Gray.
Lowell (Dr. C. W. Swan); Cambridge and Medford (F. S. Collins). July-Sept. Int. from Trop. Am. Both type and var. apparently naturalized; have been common for several years near Boston.

C. fœtidum, L.
Chelmsford, wool-waste (Dr. C. W. Swan). Aug. Adv. from Trop. Am.

"Stems herbaceous, erect, sulcate-striate, mostly simple; leaves petioled, subpatulous, oblong, sinuate-pinnatifid with obtuse lobes, thin, smoothish, glaucous green on both sides; racemes divaricately-subdichotomous, loose, leafless; fruiting calyx not closed, dentate-carinate, larger than the obtusish, channeled, smooth and somewhat shining seed." DC. Prodr. XIII, 2, 76.

C. GLAUCUM, L. OAK-LEAVED GOOSEFOOT.
Lowell, Somerville, Cambridge, et al.; abundant in waste places. June-Aug. Nat. from Eu.

C. rubrum. (Blitum maritimum, Nutt., Man.) COAST BLITE.
Salt marshes, not uncommon. July-Aug.

C. capitatum, Wats. (Blitum capitatum, L., Man.) STRAWBERRY BLITE.
Lowell, near woollen mills (Dr. C. W. Swan). June. Adv. from farther West.

C. virgatum, Wats.
Cambridge, two plants, 1885 (Walter Deane; specimen in herb. of). Adv. from Eu.
"Leaves oblong-triangular, somewhat hastate, deeply dentate, glomerules all axillary; fruiting calyx berry-like; seeds with obtuse or sometimes channeled margin." Koch, Syn. Fl. Germ.

ATRIPLEX, Tourn.

A. patula, L., var. **hastata,** Gray.
Salt marshes, common. July-Sept.

A. patula, L., var. **littoralis,** Gray.
Salt marshes, less common than the preceding. July-Sept.

A. arenaria, Nutt.
Cambridge (C. E. Perkins). Aug.-Sept.

A. bracteosa, Wats.
N. Chelmsford, wool-waste (Rev. W. P. Alcott). Adv. from Cal.
"Rather stout, suberect with spreading, flexuous branches, 2 or 3 feet high, mealy; leaves thin, sessile, lanceolate, very acute or acuminate, ½ to 1 inch long, acutely sinuate-dentate or the uppermost entire; staminate flowers in dense clusters in a naked terminal simple or compound spike; calyx deeply 5-cleft; fruiting bracts in small axillary clusters, cuneate-orbicular, 1 to 1½ lines broad, the upper rounded margin irregularly gash-toothed; the sides often somewhat muricate; seed less than half a line broad." Bot. Cal.

SALICORNIA, Tourn.

S. herbacea, L. SAMPHIRE.
Salt marshes, very common. Aug.-Oct.

S. mucronata, Bigel. (S. Virginica, L., Man.)
Salt marshes, rather common. Sept.-Oct.

SUÆDA, Forsk.

S. linearis, Torr., var. **ramosa,** Wats. (S. maritima, Dumort., Man.)
Salt marshes, very common. Aug.

SALSOLA, L.

S. Kali, L. SALTWORT.
Not uncommon in sandy soil near tide-water. Aug.

AMARANTACEÆ. AMARANTH FAMILY.

AMARANTUS, Tourn.

A. paniculatus, L.
Wakefield, Woburn and Cambridge (F. S. Collins). July-Sept. Adv. from Trop. Am.

A. RETROFLEXUS, L. PIGWEED.
Very common. July-Sept. Nat. from the South and West.

A. CHLOROSTACHYS, Willd. (A. retroflexus, L., var. chlorostachys, Man.)
Not uncommon. July-Sept. Nat. from the South and West.

A. ALBUS, L.
Common. July-Sept. Nat. from the South and West.

A. BLITUM, L.
Malden, Medford and Concord (F. S. Collins). Probably not uncommon, but generally confounded with the preceding species. July-Sept. Nat. from Eu.
"Flowers 3-parted, the axillary clusters roundish, the terminal disposed in naked spikes; stems diffuse, ascending, smooth; leaves ovate, subrhomboidal, very obtuse, retuse; bracts shorter than the flowers; capsule roundish ovate." Koch, Syn. Fl. Germ.

A. græcizans, L.
Lowell, "dump" (Dr. C. W. Swan). Aug. Adv. from S. Am.
"Like A. Blitum, but more slender, with lanceolate, obtuse leaves." DC. Prodr. XIII. 2. 263.

A. SPINOSUS, L. THORNY AMARANTH.
Lowell, waste-grounds (Dr. C. W. Swan); Malden, introduced in cotton-waste (F. S. Collins). Aug.-Sept. Nat. from Trop. Am.

A. Palmeri, Wats.
Malden, abundant in cotton-waste, 1886; appearing more sparingly in 1887 (F. S. Collins). Sept. Adv. from the Southwest.
"Diœcious, rather stout, erect, 2 or 3 feet high, branching, somewhat pubescent above or glabrate; leaves oblong-rhomboid, an inch or two long and about equalling the petiole, the upper linear

lanceolate; flowers in close elongated linear spikes leafy at the base; bracts solitary, mostly twice longer than the flowers, spreading, subulate and rigid, narrowed into a short awn; sepals of fertile flowers 1 to 1½ lines long. oblong and somewhat broader above. obtuse or retuse, two or three usually slightly larger and more acute or setaceously apiculate, distinct or nearly so; stigmas usually 2; seed circular, half a line broad." Bot. Cal.

ACNIDA, L.

A. cannabina, L.
Salt marshes. Aug.-Sept.

A. rhyssocarpa, Moq.
Salt marshes. Appears to be the more common species. Aug.-Sept.

" Fertile inflorescence very naked; the bracts not half the length of the fleshy utricle, the angles of which are not rarely rugose-tuberculated; stigmas comparatively short and slender-subulate (in A. cannabina very long and filiform, almost plumosely hairy)" A. Gray, in Am. Nat.

POLYGONACEÆ. BUCKWHEAT FAMILY.

POLYGONUM, L.

P. Bistorta, L.
Malden, a weed in and about gardens (Mrs. N. M. Hunnewell). May-June. Adv. from Eu.

" Stems usually a foot or two high; leaves few, the radical ones on long petioles, oblong-lanceolate to linear, acute at each end, 2 to 8 inches long, the cauline much reduced, mostly obtuse at base and sessile upon the sheath; the margin often slightly revolute; sheaths elongated; flowers 1½ to 2½ lines long, rose-colored to white, on slender pedicels, in very dense ovate to oblong spikes ½ to 1½ inches long and usually long-pedunculate; bracts ovate, acuminate; stamens and styles exserted; akene 1½ lines long, smooth and shining." Bot. Cal.

P. ORIENTALE, L. PRINCE'S FEATHER.
Sparingly spontaneous. July-Aug. Nat. from India.

P. Careyi, Olney.
Lowell, Bedford, Malden, et al. Not very common. July-Sept.

P. Pennsylvanicum, L.
Common. July-Sept.

P. incarnatum, Ell.
Lowell, Malden, Waltham, et al. Not very common. July-Sept

P. PERSICARIA, L. LADY'S THUMB.
Very common. July-Sept. Nat. from Eu.

P. Hydropiper, L. SMARTWEED. WATER-PEPPER.
Common. July-Sept.

P. acre, HBK. WATER SMARTWEED.
Medford (G. E. Davenport); Weston (F. S. Collins); Framingham (Rev. J. H. Temple); Concord (Minot Pratt); Dunstable (Dr. C. W. Swan). Not very common. July-Sept.

P. hydropiperoides, Michx. MILD WATER-PEPPER.
Rather common. July-Sept.

P. Mulhenbergii, Wats. (P. amphibium, L., var. terrestre, Willd., Man.)
Rather common. Aug.-Sept.

P. Hartwrightii, Gray.
Chelmsford (Dr. C. W. Swan); Fresh Pond, Cambridge (Dr. W. G. Farlow). Not common. Aug-Sept.
"Perennial, closely allied to the two preceding species, (P. amphibium and P. Muhlenbergii) growing usually in mud, the ascending stems rooting at base and very leafy; differing from the form of P. (amphibium?) growing in like localities, by being more or less rough-hairy, at least on the sheaths and bracts, the former ciliate and often with abruptly spreading foliaceous borders; leaves rather narrow, 2 to 7 inches long, on very short petioles, adnate to the middle of the sheath." Bot. Cal.

P. articulatum, L. JOINTWEED.
Common. Aug.-Sept.

P. aviculare, L. KNOTGRASS.
Very common. July-Sept.

P. erectum, L. (P. aviculare, L., var. erectum, Roth. Man.)
Not uncommon. July-Sept.

P. ramosissimum, Michx.
Watertown (C. E. Perkins); Medford (C. E. Faxon); So. Natick (F. S. Collins). Scarce, except along salt marshes. July-Sept.

P. tenue, Michx.
Frequent. Aug.-Sept.

P. arifolium, L. HALBERD-LEAVED TEAR-THUMB.
Not reported from the northern and northwestern towns, but rather common elsewhere in the county. Aug.-Sept.

P. sagittatum, L. ARROW-LEAVED TEAR-THUMB.
Common. Aug.-Sept.

P. CONVOLVULUS, L. BLACK BINDWEED.
Common. Aug.-Sept. Nat. from Eu.

P. cilinode, Michx.
Ashby (Dr. C. W. Swan); Waltham List; Newton (C. J. Sprague). Scarce.

P. dumetorum, L., var. **scandens,** Gray. Climbing False Buckwheat.
Common. Aug.-Sept.

FAGOPYRUM, Tourn.

F. ESCULENTUM, Moench. BUCKWHEAT.
Common. June-Aug. Nat. from Eu.

F. Tataricum, Gaertn. INDIA WHEAT.
Somerville (C. E. Perkins); Cambridge (*fide* specimen in herb. B. S. N. H.) July-Aug. Adv. from Asia. For description, see Wood's Bot. & Fl.

RUMEX, L.

R. PATIENTIA, L. PATIENCE DOCK.
Weston and Wakefield (F. S. Collins). July-Aug. Nat. from Eu.

R. Brittanica, L. (R. orbicularus, Gray, Man.) GREAT WATER-DOCK.
Somerville, Malden, Concord, Lowell, et al. Not uncommon. Aug.-Sept.

R. verticillatus, L. SWAMP DOCK.
Tewksbury (B. D. Greene, specimen in herb. B. S. N. H.); Somerville and Belmont (C. E. Perkins); Waltham List. June-July.

R. CRISPUS, L. CURLED DOCK. YELLOW DOCK.
Very common. June-July. Nat. from Eu.

R. OBTUSIFOLIUS, L. BITTER DOCK.
Common. June-Aug. Nat. from Eu.

R. CRISPUS, L. × OBTUSIFOLIUS, L.
Malden (F. S. Collins); Medford (L. L. Dame). June-Aug.

R. ACETOSELLA, L. SORREL.
Very common. May-June. Nat. from Eu.

CHORIZANTHE, R. Br.

"Involucre 1-3 flowered, sessile, tubular, coriaceous or chartaceous, more or less reticulated or corrugated, 3-6 angled or—costate and 3-6 toothed or—cleft, the teeth cuspidate or rigidly awned. Flowers pedicellate or nearly sessile, included in the involucre or rarely exserted, 6 parted or cleft. Stamens 9, rarely 3 or 6. Bractlets minute or usually obsolete. Ovary glabrous and akene triangular. Low, dichotomously branched annuals, with usually rosulate, radical leaves, and ternate bracts." Bot. Cal.

C. pungens, Benth.
N. Chelmsford, wool-waste (Rev. W. P. Alcott; specimen in herb. of). Adv. from Cal.

"Usually slender and more or less decumbent or at first erect, villous-pubescent; stems often a foot long or more, leafy; leaves spatulate or oblanceolate, about an inch long, mostly opposite; bracts similar, narrower, awned at the apex; heads small; involucres 1½ to 2 lines long, unequally toothed (the alternate teeth smaller), usually margined; teeth strongly uncinate; flowers very shortly pedicelled, narrowed at base, 1½ lines long, glabrous or often villous on the nerves, shortly cleft; segments equal, oblong, entire; filaments more or less adnate to the lower part of the tube. Bot. Cal.

LAURACEÆ. LAUREL FAMILY.

SASSAFRAS, Nees.

S. officinale, Nees.
Common. Apr.

LINDERA, Thunb.

L. Benzoin, Meisn. SPICE-BUSH. BENJAMIN BUSH.
Common. Apr.

THYMELEACEÆ. MEZEREUM FAMILY.

DIRCA, L.

D. palustris, L. LEATHERWOOD. MOOSEWOOD. WICOPY.
Townsend, rare (Miss H. E. Haynes). Concord, introduced from Vt. by Minot Pratt. Apr.

DAPHNE, L.

D. MEZEREUM, L. DAPHNE.
Medford, persistent in two localities for many years, without much tendency to spread. Apr. Nat. from Eu. For description see Wood's Bot. & Fl.

SANTALACEÆ. SANDALWOOD FAMILY.

COMANDRA, Nutt.

C. umbellata, Nutt. BASTARD TOAD-FLAX.
Common. May-June.

CERATOPHYLLACEÆ. HORNWORT FAMILY.

CERATOPHYLLUM, L.

C. demersum, L. HORNWORT.
Cambridge (Rev. Thos. Morong); Newton (F. S. Collins).

C. demersum, L., var. **echinatum,** Gray.
Cambridge, specimen in herb. B. S. N. H.

CALLITRICHACEÆ. WATER-STARWORT FAMILY.

CALLITRICHE. L.

C. verna, L. WATER STARWORT.
Common. May-July.

C. heterophylla, Pursh.
Ashland (Rev. Thos. Morong); Malden (H. A. Young); Spot Pond (Wm. Boott). May-July.

PODOSTEMACEÆ. RIVER-WEED FAMILY.

PODOSTEMON, Michx.

P. ceratophyllus, Michx. RIVER-WEED.
In the Charles River, So. Natick (Edwin Faxon). July-Sept.

EUPHORBIACEÆ. SPURGE FAMILY.

EUPHORBIA, L.

E. maculata, L.
Very common. July-Sept.

E. hypericifolia, L.
Lowell (Dr. C. W. Swan); Concord (Minot Pratt); Framingham (Rev. J. H. Temple). Rare. July-Sept.

E. corollata, L. FLOWERING SPURGE.
Lowell, "dump" (Dr. C. W. Swan). July-Sept. Adv. from the West.

E. Esula, L.
Somerville (C. E. Perkins); Waltham List. Rare. June. Adv. from Eu.

E. CYPARISSIAS, L. CYPRESS SPURGE.
Common. May-June. Nat. from Eu.

ACALYPHA, L.

A. Virginica, L. THREE-SEEDED MERCURY.
Common. July-Sept.

URTICACEÆ. NETTLE FAMILY.

ULMUS, L.

U. fulva, Michx. SLIPPERY ELM. RED ELM.
Concord (Walter Deane). Rare. Apr.

U. Americana, L. AMERICAN ELM. WHITE ELM.
Common. Apr.

CELTIS, Tourn.

C. occidentalis, L, including var. **crassifolia.** SUGARBERRY. HACKBERRY.
Lowell, Emerson's Trees and Shrubs of Mass. Apr.-May.

MORUS, Tourn.

M. rubra, L. RED MULBERRY.
Sparingly spontaneous. May.

M. ALBA, L. WHITE MULBERRY.
Occasional; N. Billerica, along the old Middlesex Canal, among wild shrubs, itself shrubby (Dr. C. W. Swan); Concord, several trees (Minot Pratt). May. Nat. from Eu.

URTICA, Tourn.

U. gracilis, Ait. NETTLE.
Common. July-Aug.

URTICA DIOICA, L.
Fence-row. Watertown (Walter Deane.) Nat. from Eu.

U. URENS, L.
Malden (H. A. Young); Concord (Minot Pratt). Scarce. July-Aug. Nat. from Eu.

LAPORTEA, Gaudich.

L. Canadensis, Gaudich. WOOD NETTLE.
Natick (Austin Bacon); Townsend (Dr. C. W. Swan); Belmont (L. L. Dame). July-Aug.

PILEA, Lindl.

P. pumila, Gray. RICHWEED. CLEARWEED.
Rather common. July-Sept.

BŒHMERIA, Jacq.

B. cylindrica, Willd. FALSE NETTLE.
Common. July-Aug.

PARIETARIA, Tourn.

*P. Pennsylvanica, Muhl. PELLITORY.
Concord (Minot Pratt). This plant is out of its usual range, and may possibly have been introduced by accident with other plants. June-Aug.

CANNABIS, Tourn.

C. sativa, L. HEMP.
Rather common. July-Aug. Nat. from Eu.

HUMULUS, L.

H. Lupulus, L. HOP.
Sparingly naturalized from the West. July.

PLATANACEÆ. PLANE-TREE FAMILY.

PLATANUS, L.

P. occidentalis, L. SYCAMORE. BUTTONWOOD.
Rather common. May-June.

JUGLANDACEÆ. WALNUT FAMILY.

JUGLANS, L.

J. cinerea, L. BUTTERNUT.
Common. May.

CARYA, Nutt.

C. alba, Nutt. SHAGBARK HICKORY.
Common. May-June.
C. tomentosa, Nutt. MOCKER-NUT. WHITE-HEART HICKORY.
Not uncommon. May-June.
C. porcina, Nutt. PIGNUT or BROWN HICKORY.
Common. May-June.
C. amara, Nutt. BITTERNUT HICKORY.
Medford, Winchester, Belmont, et al. Less common than the other species. May-June.

CUPULIFERÆ. OAK FAMILY.

QUERCUS, L.

Q. alba, L. WHITE OAK.
Common. May.
Q. bicolor, Willd. SWAMP WHITE OAK.
Common, especially in the eastern section of the county. May.

Q. Prinus, L. CHESTNUT OAK.
Medford (L. L. Dame); Townsend (John H. Sears). Rare. May.
Q. prinoides, Willd. (Q. Prinus, L., var. humilis, Marsh.)
Common. May.
Q. ilicifolia, Wang. BEAR or SCRUB OAK.
Common. Apr.-May.
Q. coccinea, Wang. SCARLET OAK.
Rather common. May.
Q. tinctoria, Bartr. YELLOW or BLACK OAK.
Common. May.
Q. rubra, L. RED OAK.
Common. May.

CASTANEA, Tourn.

C. vulgaris, Lam., var. AMERICANA, A. DC. (C. vesca, L., var. Americana, Michx., Man.) CHESTNUT.
Common, especially in the western part of the county. July.

FAGUS, Tourn.

F. ferruginea, Ait. BEECH.
Rather common. May-June.

CORYLUS, Tourn.

C. Americana, Walt. HAZLENUT.
Common. Mch.-April.
C. rostrata, Ait. BEAKED HAZLENUT.
Widely distributed, but not so common as the preceding species. Mch.-April.

OSTRYA, Mich.

O. Virginica, Willd. HOP HORNBEAM. LEVER-WOOD. IRON WOOD.
Rather common. Apr.-May.

CARPINUS, L.

C. Caroliniana, Walt. (C. Americana, Michx., Man.) HORN-BEAM. IRON WOOD. BLUE BEECH.
Generally distributed, not uncommon. May.

MYRICACEÆ. SWEET-GALE FAMILY.

MYRICA. L.

M. Gale, L. SWEET GALE.
Rather common. Apr.
M. cerifera, L. BAYBERRY. WAX-MYRTLE.
Common. May-June.

COMPTONIA, Soland.

C. asplenifolia, Ait. SWEET FERN.
Common. Apr.-May.

BETULACEÆ. BIRCH FAMILY.

BETULA, Tourn.

B. lenta, L. CHERRY BIRCH. SWEET or BLACK BIRCH.
Widely distributed. Frequent. Apr.-May.
B. lutea, Michx, f. YELLOW BIRCH.
Not as common as the preceding species. Apr.-May.
B. alba, L. var. populifolia, Spach. AMERICAN WHITE BIRCH. GRAY BIRCH.
Common. May.
B. papyracea, Ait. WHITE BIRCH. PAPER or CANOE BIRCH.
Rare eastward, but common in other sections of the county. May-June.
B. papyracea, Ait., var. minor, Tuckerm. DWARF CANOE BIRCH.
A clump of trees 6 or 7 feet high, growing in a swamp in Lexington, 1875 (Minot Pratt).
B. nigra, L. RIVER BIRCH. RED BIRCH.
Native only in the Merrimac River Valley.

ALNUS, Tourn.

A. incana, Willd. BLACK ALDER. SPECKLED or HOARY ALDER.
Common. Mch.-Apr.
A. serrulata, Willd. SMOOTH ALDER.
Less common than the preceding species. Mch.-Apr.

SALICACEÆ. WILLOW FAMILY.

SALIX, Tourn.

S. tristis, Ait. DWARF GRAY WILLOW.
Not uncommon. Apr.-May.
S. humilis, Marsh. PRAIRIE WILLOW.
Rather common. "The intermediate character of this species, as between S. tristis and S. discolor was long ago pointed out by Mr. Carey. The confusing forms appear to be hybrids."—M. S. Bebb.
S. discolor, Muhl. GLAUCOUS WILLOW.
Common. Forms with anthers transformed to ovaries, occasional. Apr.-May.

S. sericea, Marsh. SILKY WILLOW.
Medford (G. E. Davenport); Arlington (Wm. Boott); Hopkinton (L. L. Dame). Not common. May.

S. petiolaris, Smith. PETIOLED WILLOW.
Medford (G. E. Davenport); Winchester (L. L. Dame); Cambridge (L. H. Bailey, Jr.), et al. Apr.-May.

S. PURPUREA, L. PURPLE WILLOW.
Arlington (Wm. Boott); Medford (L. L. Dame). Apr.-May. A native of the Old World, sparingly naturalized.

S. viminalis, L. BASKET OSIER.
West Medford (C. E. Perkins). Adv. from Eu. May.

S. cordata, Muhl. HEART-LEAVED WILLOW.
Widely distributed, but not common. Apr.-May.

S. cordata × sericea, Bebb. ("S. myricoides, Muhl.! not the S. cordata, var. myricoides, of Gray, Man.")—M. S. Bebb.
Near Fresh Pond, Cambridge (L. H. Bailey, Jr.)

S. rostrata, Rich. (S. livida, Wahl., var. occidentalis, Gray.) LIVID WILLOW.
Common. Apr.-May. Mr. Bebb ranks this willow as a sub-species of S. livida.

S. lucida, Muhl. SHINING WILLOW.
Rather common. May.

S. nigra, Marsh. BLACK WILLOW.
Cambridge, Medford, Winchester, et al. Occasional.

S. nigra, Marsh, var. **falcata,** Gray.
Cambridge (Dr. C. W. Swan); Concord (Walter Deane).

S. FRAGILIS, L. BRITTLE WILLOW.
Hybrids in which S. fragilis predominates are not uncommon; a Medford specimen Mr. Bebb pronounces "almost pure *fragilis.*" May.

S. ALBA, L., var. VITELLINA, Koch.
Chelmsford, Medford, Natick, et al. Not uncommon. May. "The typical S. alba is extremely rare in the United States, and what Anderson, Wimmer, and the German botanists generally regard as genuine S. fragilis I have not seen at all; but the var. vitellina is very common, as are also a host of hybrid forms between alba and fragilis, representing S. viridis, Fr., S. Russelliana, Sm., &c. These hybrids, perplexing enough in themselves, are rendered still more inextricable, with us, by a further cross — by no means rare — with native lucida,"—M. S. Bebb.

S. myrtilloides, L. MYRTLE WILLOW.
Cambridge, Belmont, Wakefield, Chelmsford, et al. Rather common. May.

POPULUS, Tourn.

P. tremuloides, Michx. AMERICAN ASPEN.
Common. Found with polygamous flowers, Medford, 1878 (G. E. Davenport). Apr.

P. grandidentata, Michx. LARGE-TOOTHED ASPEN.
Common. Apr.

P. balsamifera, L., var. **candicans,** Gray. BALM OF GILEAD.
Common. Possibly introduced from farther North. May.

P. dilatata, Ait. LOMBARDY POPLAR.
Formerly extensively cultivated, and occasionally spontaneous, but apparently dying out. Adv. from Eu.

P. ALBA, L. ABELE. WHITE-POPLAR.
Occasional; spreading extensively by root. Nat. from Eu.

CONIFERÆ. PINE FAMILY.

PINUS, Tourn.

P. rigida, Mill. PITCH PINE.
Common. May.

P. resinosa, Ait. NORWAY PINE. RED PINE.
Generally distributed, but not common. May.

P. Strobus, L. WHITE PINE.
Common. June.

PICEA, Link.

P. nigra, Link. (Abies nigra of Man.) BLACK SPRUCE.
Rather scarce, especially eastward. May.

TSUGA, Endl.

T. Canadensis, Carriere. (Abies Canadensis, Michx., Man.) HEMLOCK.
Rather common. June.

ABIES, Tourn.

A. balsamea, Marsh. FIR BALSAM.
Ashby, near Mt. Watatic (Dr. C. W. Swan); Concord, introduced by Minot Pratt). Rare. May.

LARIX, Tourn.

L. Americana, Michx. AMERICAN LARCH. BLACK L. TAMARACK. HACKMATAC.
Widely distributed; not uncommon. Apr.-May.

THUYA, Tourn.

T. occidentalis, L. AMERICAN ARBOR-VITÆ.
Concord, introduced by Minot Pratt. May-June.

CHAMÆCYPARIS, Spach.

C. sphæroidea, Spach. (Cupressus thyoides, L., Man.) WHITE CEDAR.
Reading, Bedford, Natick, Hopkinton, et al. Not very common. May.

JUNIPERUS, L.

J. communis, L. JUNIPER.
Common. May-June.

J. Virginiana, L. RED CEDAR. SAVIN.
Common. May.

TAXUS, Tourn.

T. baccata, L., var. **Canadensis,** Gray.
AMERICAN YEW. GROUND HEMLOCK.
Common in Ashby (L. L. Dame); very rare elsewhere. Apr.

ENDOGENS.

ARACEÆ. ARUM FAMILY.

ARISÆMA, Mart.

A. triphyllum, Torr. INDIAN TURNIP. JACK-IN-THE-PULPIT.
Common. May-June.

PELTANDRA, Raf.

P. undulata, Raf. (P. Virginica, Raf., Man., in part).
Not uncommon. June.
Again separated, in Engler's Monograph, from P. Virginica, Raf. The most obvious distinction is to be found in the spadix, the pistillate part of which, in P. undulata, is from 1–4 to 1–5, while in P. Virginica, it is 2–3 the length of the staminate part. All the county specimens examined seem to be P. undulata.

CALLA, L.

C. palustris, L. WATER ARUM.
Widely distributed, but not very common. May-June.

7

SYMPLOCARPUS, Salisb.

S. fœtidus, Salisb. SKUNK CABBAGE.
Common. Mar.-Apr.

ACORUS, L.

A. Calamus, L. SWEET FLAG.
Common. June. Apparently native.

LEMNACEÆ. DUCKWEED FAMILY.

LEMNA, L.

L. trisculca, L.
Fresh Pond, Cambridge (Wm. Boott). Rare.
L. minor, L. DUCKWEED.
Common.

SPEIRODELA, Schleid.

S. polyrrhiza, Schleid. (Lemna polyrrhiza, L., Man.)
Common.

TYPHACEÆ. CAT-TAIL FAMILY.

TYPHA, Tourn.

T. latifolia, L. CAT-TAIL.
Common. June-July.
T. angustifolia, L. SMALL CAT-TAIL.
Winchester and Arlington (C. E. Perkins); Medford (G. E. Davenport). Common in these localities, but not reported elsewhere. June-July.

SPARGANIUM, Tourn.

S. eurycarpum, Engelm. BUR-REED.
Rather common.
S. androcladum, Morong in Torrey Bulletin, March, 1888. (S. simplex, Huds., var. androcladum, Engelm.) Whitehall Pond, Hopkinton (Rev. Thos. Morong); Medford (Wm. Boott); Waltham (C. E. Perkins); et al.
S. androcladum, Morong, var. **fluctuans,** Morong in Torrey Bulletin, March, 1888. (S. simplex, Huds., var. fluitans, Engelm., Man.) A form closely approaching this variety was collected at Whitehall Pond, Hopkinton, by Dr. C. W. Swan.
S. simplex, Huds.
Common. The form known as var. **Nuttallii,** Engelm., Man., likewise common.

S. minimum, Fries.
Ashland (Rev. Thos. Morong).

NAIADACEÆ. PONDWEED FAMILY.

NAIAS, L.

N. flexilis, Rostk.
Common.

N. flexilis, Rostk., var. robusta, Morong.
Concord River, abundant (Walter Deane); Natick (Rev. Thos. Morong; specimen in herb. of).
"Stem stout, few leaved, sparsely branching, elongated; leaves linear, 1½—2mm. broad and 10—15mm. long, flat, abruptly acute. I have found it rising to the surface in still ponds, in water 4 to 6 feet deep." Morong in Bot. Gaz. The usual fruit of the type.

N. Indica, Cham., var. gracillima, Braun.
Woburn (Wm. Boott); Spot Pond, Stoneham, and Ashland (Rev. Thos. Morong). Rare.

ZANNICHELLIA, Mich.

Z. palustris, L. HORNED PONDWEED.
Very abundant in Mystic Pond and River (Wm. Boott). June-July.

ZOSTERA, L.

Z. marina, L. EEL-GRASS.
Very common in salt water. July-Aug.

RUPPIA, L.

R. maritima, L. DITCH-GRASS.
Common in brackish water. June-July.

POTAMOGETON, Tourn.

P. natans, L.
Common.

P. Oakesianus, Robbins.
Mystic Pond, Medford, and Horn Pond, Woburn (Wm. Boott); Natick (Rev. Thos. Morong).

P. Claytonii, Tuck.
Common.

P. Vaseyi, Robbins.
Mystic Pond, Medford (Wm. Boott); Spot Pond, Stoneham (Rev. Thos. Morong).

P. Spirillus, Tuck.
Westford (Dr. C. W. Swan); Woburn, Winchester, Medford and Billerica (Wm. Boott); Concord (E. S. Hoar).

P. hybridus, Michx.
Winchester; Silver Lake, Wilmington; and Round Pond, Woburn (Wm. Boott); Bedford (Dr. C. W. Swan); Medford, clay-pits (F. S. Collins).

P. lonchites, Tuck.
Winchester (Rev. Thos. Morong).

P. pulcher, Tuck.
Spot Pond, Stoneham, and Ashland (Rev. Thos. Morong); Fresh Pond, Cambridge (E. Tuckerman); Concord (Walter Deane).

P. amplifolius, Tuck.
Mystic Pond, Medford (Wm. Boott); Fresh Pond, Cambridge (Rev. Thos. Morong); Concord (Walter Deane); Townsend and Bedford (Dr. C. W. Swan).

P. gramineus, L.
Common.

P. gramineus, L., var. **spathulæformis,** Robbins.
Mystic Pond (Rev. Thos. Morong).

P. gramineus, L., var. **maximus,** Morong.
So. Natick (Rev. Thos. Morong).
"This variety generally occurs in swift currents, and differs from the type in usually having all the parts much elongated, stems 5 to 10 ft. in length, and the sessile or petiolate submerged leaves 3-7 lines wide by 2-5 inches long, and 7-10 nerved." Morong *in litt.*

P. lucens, L.
Fresh Pond (Rev. Thos. Morong); Winchester (L. L. Dame).

P. Zizii, Mert. & Koch. (P. lucens, L., var. minor, Nolte, Man.)
Fresh Pond (Rev. Thos. Morong).

P. prælongus, Wulf.
Fresh Pond (John Robinson, et al.)

P. perfoliatus, L.
Mystic Pond (Wm. Boott); Fresh Pond (Dr. C. W. Swan).

P. Mysticus, Morong.
Mystic Pond, Medford (Rev. Thos. Morong).
"The whole plant very slender; stems irregularly branching from a creeping rootstock, nearly filiform, terete. 1-3 ft. high; leaves all submerged, scattered, entire, oblong-linear, $\frac{1}{2}$-1$\frac{1}{2}$ inches long, and 2 or 3 lines wide, 5-7 nerved, finely undulate, obtuse or bluntly pointed at the apex, abruptly narrowing at the base, and sessile or partly clasping; stipules free, obtuse, about 6 lines long, mostly deciduous but often persistent, and closely sheathing the stem; spikes few, capitate, 4-6 flowered, on erect peduncles from 1-2 inches long. With the habit of P. perfoliatus, but scarcely $\frac{1}{3}$ as stout in any of its parts." Morong in Bot. Gaz., Vol. V., No. 5.

P. crispus, L.
  Spy Pond, Arlington; probably introduced (C. E. Faxon).
P. zosteræfolius, Schum. (P. compressus, L., Man.)
  Cambridge and Natick (Rev. Thos. Morong).
P. obtusifolius, Mert. & Koch.
  Natick (Rev. Thos. Morong, specimen in herb. of).
P. pauciflorus, Pursh.
  Medford (Wm. Boott); Cambridge (Rev. Thos. Morong).
P. pusillus, L.
  Common.
P. pusillus, L., var. **polyphyllus,** Morong.
  So. Natick and Fresh Pond, Cambridge (Rev. Thos. Morong).
  "A dwarf form, 3-5 inches high, divaricately branching from the base, and very leafy throughout; leaves very obtuse, not cuspidate, 3-nerved; non-flowering, but abundantly provided with propagating buds which are formed on the thickened and hardened ends of the branches, and closely invested by imbricated leaves." Morong in Bot. Gaz., Vol. V., No. 5.
P. pusillus, L., var, **tenuissimus,** Mert. & Koch.
  Natick (Rev. Thos. Morong).
P. gemmiparus, Robbins, (P. pusillus, L., var. gemmiparus, Man.)
  Mystic Pond (E. Tuckerman); So. Natick (Rev. Thos. Morong).
P. Robbinsii, Oakes.
  Fresh Pond (E. Tuckerman); Medford (Wm. Boott); Concord (Walter Deane).

TRIGLOCHIN, L.

T. maritimum, L. ARROW-GRASS.
  Common in salt marshes. June-July.

SCHEUCHZERIA, L.

S. palustris, L.
  Ashby (W. H. Manning); Tewksbury (B. D. Greene). June-July.

ALISMACEÆ. WATER-PLANTAIN FAMILY.

ALISMA, L.

A. Plantago, L., var. **Americanum,** Gray. WATER-PLANTAIN.
  Rather common. June-July.

ECHINODORUS, Rich., Engelm.

E. parvulus, Engelm.
  Cambridge (James); Winter Pond, Winchester (Dr. C. W. Swan). July-Aug.

SAGITTARIA, L.

S. variabilis, Engelm. ARROW-HEAD.
Very common; including vars. **obtusa** and **latifolia**, Engelm., Malden (H. L. Moody); var. **hastata**, Engelm., Cambridge (G. E. Davenport); var. **diversifolia**, Engelm., Lowell (W. H. Manning); and var. **angustifolia**, Lowell (L. L. Dame). June-Aug.

S. calycina, Engelm.
Medford (Wm. Boott, specimen in herb. of). July.

S. graminea, Michx.
Cambridge and Arlington (C. E. Perkins); Framingham and Natick (Rev. Thos. Morong); Medford (W. H. Manning). Aug.

S. graminea, Michx., var. **fluitans**, Engelm.
So. Natick (Rev. Thos. Morong; specimen in herb. of). Aug.

S. natans, Michx.
Westford (C. E. Perkins); Littleton (W. H. Manning).

HYDROCHARIDACEÆ. FROG'S-BIT FAMILY.

ANACHARIS, Rich.

A. Canadensis, Planch. WATER-WEED.
Somerville (Warner Bailey); Arlington (Wm. Boott); Fresh Pond, Cambridge (Rev. Thos. Morong). July.

VALLISNERIA, Mich.

V. spiralis, L. TAPE-GRASS. EEL-GRASS.
Common. July-Aug.

ORCHIDACEÆ. ORCHID FAMILY.

ORCHIS, L.

*****O. spectabilis**, L. SHOWY ORCHIS.
Cambridge (T. W. Harris in Hovey's Mag., VI., 245). Concord, introduced from Vermont by Minot Pratt. May. Very rare.

HABENARIA, Willd., R. Br.

H. tridentata, Hook.
Common. June-Aug.

H. virescens, Spreng.
Not uncommon. June-July.

H. dilatata, Gray.
Lexington (Baldwin's Orchids of N. E.); Reading and Stoneham (Wm. Boott, specimen in herb. of). The Concord plant was introduced by Minot Pratt. June-July.

H. obtusata, Richards.
   Concord, introduced from Wachusett by Minot Pratt. June.
H. Hookeri, Torr.
   Ashby (W. H. Manning); Stoneham (Mrs. P. D. Richards); Hopkinton (Dr. C. W. Swan). Rare. June.
H. orbiculata, Torr.
   Generally distributed, but not common. July.
H. ciliaris, R. Br. YELLOW FRINGED-ORCHIS.
   Lexington, 1862 (Wm. Boott), specimen in the Gray herb. Very rare. July-Sept.
H. blephariglottis, Hook. WHITE FRINGED-ORCHIS.
   Generally distributed, but rare. July-Aug.
H. blephariglottis, Hook., var. holopetala, Gray.
   Littleton (W. H. Manning). July-Aug.
H. lacera, R. Br. RAGGED FRINGED-ORCHIS.
   Common. July-Aug.
H. psycodes, Gray. PURPLE FRINGED-ORCHIS.
   Common. A form at Malden with white flowers (F. S. Collins). July-Aug.
H. fimbriata, R. Br. LARGER PURPLE FRINGED-ORCHIS.
   Generally distributed, but not common. June-July.

GOODYERA, R. Br.

G. repens, R. Br. RATTLESNAKE PLANTAIN.
   Not uncommon. Aug.
G. pubescens, R. Br. RATTLESNAKE PLANTAIN.
   Rather common. July-Aug.

SPIRANTHES, Rich.

S. latifolia, Torr.
   Groton (Miss H. E. Haynes); N. Reading (W. H. Manning). Rare. June.
S. cernua, Rich. LADIES' TRESSES.
   Common. Aug.-Oct.
S. graminea, Lindl., var. Walteri, Gray.
   Townsend (Miss H. E. Haynes); Billerica (W. H. Manning). Rare. July-Aug.
S. gracilis, Bigel. SLENDER LADIES' TRESSES.
   Common. July-Oct.

ARETHUSA, Gronov.

A. bulbosa, L.
   Frequent. May.

POGONIA, Juss.

P. ophioglossoides, Nutt.
Common. June-July.

P. verticillata, Nutt.
Ashby (W. H. Manning); Tewksbury (Rev. H. C. Dugauno); Lowell (Dr. F. Nickerson); Billerica (Baldwin's Orchids of N. E.). Rare. May-June.

CALOPOGON, R. Br.

C. pulchellus, R. Br.
Common. June-July.

MICROSTYLIS, Nutt.

*****M. ophioglossoides,** Nutt.
Wilmington (E. H. Hitchings and Miss M. E. Carter). Very rare. July.

LIPARIS, Rich.

*****L. liliifolia,** Rich. TWAYBLADE.
Cambridge, (Bigelow's Fl. Bost.); Natick (Austin Bacon). Very rare. June.

L. Loeselii, Rich.
Tewksbury and Chelmsford (Dr. C. W. Swan); Woburn (Miss M. E. Carter); Ashland (Rev. Thos. Morong); Belmont (C. E. Perkins); W. Medford (Mrs. P. D. Richards). Rare. June.

CORALLORHIZA, Hall.

C. innata, R. Br.
Stoneham and Medford (C. E. Perkins); Bedford (C. W. Jenks); Natick (Baldwin's Orchids of N. E.); Tewksbury (B. D. Greene). May-June.

C. multiflora, Nutt.
Rather common. July-Sept.

CYPRIPEDIUM, L.

*****C. arietinum** R. Br. RAM'S-HEAD LADY'S SLIPPER.
Concord, introduced from Vermont by Minot Pratt. June.

C. parviflorum, Salisb. SMALLER YELLOW LADY'S SLIPPER.
Groton (Miss. H. E. Haynes); Westford (Baldwin's Orchids of N. E.); Acton (Walter Deane; specimen in herb. of); Concord, introduced by Minot Pratt. Very rare. June.

C. pubescens, Willd. LARGER YELLOW LADY'S SLIPPER.
Reading and Lexington (Baldwin's Orchids of N. E.); Ashland (Rev. Thos. Morong); Stowe (Walter Deane; specimen in herb. of); Concord, introduced by Minot Pratt. Very rare. June.

C. spectabile, Sw. SHOWY LADY'S SLIPPER.
Wilmington (Baldwin's Orchids of N. E.); Reading, found in a cedar swamp in 1866 (Miss Clymena Wakefield); Concord, introduced from Conway, N. H., by Minot Pratt. Very rare. July.

C. acaule, Ait. LADY'S SLIPPER.
Common. A form with white flowers reported at Ashby by F. S. Collins, and at N. Reading by John Robinson; rare. May-June.

AMARYLLIDACEÆ. AMARYLLIS FAMILY.

HYPOXYS, L.

H. erecta, L. STAR GRASS.
Common. May-Aug.

HÆMODORACEÆ. BLOODWORT FAMILY.

ALETRIS, L.

*A. farinosa, L. COLIC-ROOT. STAR-GRASS.
Framingham (Rev. J. H. Temple). Rare. July-Aug.

IRIDACEÆ. IRIS FAMILY.

IRIS, L.

I. versicolor, L. LARGER BLUE FLAG.
Common. May-June.

I. Virginica, L. SLENDER BLUE FLAG. BOSTON IRIS.
Less common than the preceding. May-June.

I. Pseudacorus, L.
Occasionally escaped, but hardly established. Int. from Eu.
"Beardless; leaves sword-shaped, lanceolate linear, about equalling the terete, many flowered stem; sepals ovate with a broad claw, the linear petals narrower and shorter than the pistils. Flowers yellow, the sepals having in place of the beard, a yellow spot, marked with very fine black or dark purple veins." Koch, Syn. Fl. Germ.

SISYRINCHIUM, L.

S. anceps, Cav. (S. Bermudiana, L., Man., in part.)
Common. May-July. A form with pure white flowers has been observed at Reading by W. H. Manning.

S. mucronatum, Michx. (S. Bermudiana, L., var. mucronatum, Man.)
Wilmington (F. S. Collins); Chelmsford (Dr. C. W. Swan). Much the less common species. May-July.

DIOSCOREACEÆ. YAM FAMILY.

DIOSCOREA, Plum.

*****D. villosa,** L. WILD YAM ROOT.
"Woods, Concord Turnpike" (Bigelow's Fl. Bost.) Not since reported. July.

SMILACEÆ. SMILAX FAMILY.

SMILAX, Tourn.

S. rotundifolia, L. GREEN BRIAR.
Common. June-July.

S. glauca, Walt.
Weston (L. L. Dame). An extreme northern station. July.

S. herbacea, L. CARRION-FLOWER.
Common. June.

*****S. herbacea,** L., var. **peduncularis,** A. DC. (var. pulverulenta, Man.)
Malden (F. S. Collins); Natick (Austin Bacon); Marlboro (Mrs. A. M. Staples). June.

LILIACEÆ. LILY FAMILY.

TRILLIUM, L.

*****T. sessile,** L.
Concord, introduced from the West by Minot Pratt. June.

*****T. recurvatum,** Beck.
Concord, introduced from Wisconsin by Minot Pratt. June.

*****T. grandiflorum,** Salisb. LARGE-FLOWERED TRILLIUM.
Concord, introduced by Minot Pratt. June.

T. erectum, L. PURPLE TRILLIUM. BIRTHROOT.
Townsend (Miss H. E. Haynes and C. W. Jenks); Ashby (L. L. Dame). Concord, introduced from Wachusett by Minot Pratt. Rare. May.

T. cernuum, L. NODDING TRILLIUM. WAKE-ROBIN.
Common. May-June.

*T. nivale, Ridd. DWARF WHITE TRILLIUM.
Concord, introduced from Illinois by Minot Pratt. Apr.-May.
T. erythrocarpum, Michx. PAINTED TRILLIUM.
Ashby, common (L. L. Dame); Groton (C. W. Jenks); Concord (Minot Pratt); Tewksbury (B. D. Greene); Framingham (Rev. J. H. Temple). Rare, except in the northern towns. May-June.

MEDEOLA, Gronov.

M. Virginica, L. INDIAN CUCUMBER-ROOT.
Common. June.

VERATRUM, Tourn.

V. viride, Ait. AMERICAN WHITE HELLEBORE. INDIAN POKE.
Common. June-July.

UVULARIA, L.

*U. grandiflora, Smith.
Concord, introduced from Vermont by Minot Pratt. May-June.
U. perfoliata, L. BELLWORT.
Rather common. May.

OAKESIA, Wats.

O. sessilifolia, Wats. (Uvularia sessilifolia, L., Man.) BELLWORT. WILD OATS.
Common. May.

STREPTOPUS, Michx.

S. amplexifolius, DC.
Ashby (Dr. C. W. Swan); Concord, introduced from Wachusett by Minot Pratt. June.

CLINTONIA, Raf.

C. borealis, Raf.
Common, especially in the northern part of the county. May-June.

CONVALLARIA, L.

C. majalis, L. LILY OF THE VALLEY.
Concord, introduced by Minot Pratt; Wakefield, apparently established (F. S. Collins). May-June. Adv. from Eu. and from the South.

SMILACINA, Desf.

S. racemosa, Desf. FALSE SPIKENARD.
Common. June.
S. stellata, Desf.
Waverly (Wm. Boott); near Fresh Pond, Cambridge (C. E. Faxon); Concord, introduced from Breed's Island by Minot Pratt. May-June.

S. trifolia, Desf.
　Occasional throughout the county, but nowhere common. May.

MAIANTHEMUM, Desf.

M. Canadense, Desf.　(Smilacina bifolia, Ker, Man.)　FALSE SOLOMON'S SEAL.
　Very common. May-June.

POLYGONATUM, Tourn.

P. biflorum, Ell.　SMALLER SOLOMON'S SEAL.
　Common. May-June.

ASPARAGUS, L.

A. OFFICINALIS, L.　GARDEN ASPARAGUS.
　Common. June-July. Nat. from Eu.

LILIUM, L.

L. Philadelphicum, L.　WILD RED LILY.　WOOD LILY.
　Common. July.
L. Canadense, L.　WILD YELLOW LILY.　MEADOW LILY.
　Common. July.
L. TIGRINUM, L.　TIGER LILY.
　Occasionally spontaneous and locally established. July-Aug. Nat. from Eu. For description, see Wood's Bot. & Fl.

ERYTHRONIUM, L.

E. Americanum, Smith.　DOG'S-TOOTH VIOLET.　YELLOW-ADDER'S-TONGUE.
　Frequent. Apr.-May.

ORNITHOGALUM, Tourn.

O. UMBELLATUM, L.　STAR-OF-BETHLEHEM.
　Not uncommon. June. Nat. from Eu.

SCILLA, L.

*S. Fraseri, Gray.
　Concord, introduced from Illinois by Minot Pratt. May.

ALLIUM, L.

A. tricoccum, Ait.　WILD LEEK.
　Framingham (Rev. J. H. Temple). Very rare. July.
A. Canadense, Kalm.　WILD GARLIC.
　Frequent. May-June.

YUCCA, L.

*Y. filamentosa, L. ADAM'S NEEDLE.
Reading, "persistent for several years on waste heaps, and spreading slowly by roots" (W. H. Manning). Adv. from the South. July.

HEMEROCALLIS, L.

H. FULVA, L. DAY LILY.
Locally established. July. Nat. from Eu.
H. FLAVA, L. YELLOW DAY LILY.
Locally established. July. Nat. from Eu.

JUNCACEÆ. RUSH FAMILY.

LUZULA, DC. WOOD-RUSH.

L. campestris, DC.
Common.

JUNCUS, L. RUSH. BOG-RUSH.

J. effusus, L. COMMON RUSH. SOFT RUSH.
Common.
***J. effusus, L., var. conglomeratus,** Gray.
Concord (H. S. Richardson).
J. filiformis, L.
Concord, (E. S. Hoar; specimen in herb. of).
J. Balticus, Dethard.
Medford (Wm. Boott).
J. marginatus, Rostk.
Common.
J. bufonius, L.
Dracut (Dr. C. W. Swan); Medford (G. E. Davenport); Somerville (C. E. Perkins); Concord (E. S. Hoar).
J. Gerardi, Loisel. BLACK GRASS.
Very common on salt marshes.
J. tenuis, Willd.
Common.
J. tenuis, Willd., var. **secundus,** Gray.
Winchester (Rev. G. P. Huntington).
J. Greenii, Oakes & Tuck.
Pepperell, Dracut and Lowell (Dr. C. W. Swan); Medford (Wm. Boott).
J. pelocarpus, E. Meyer.
Generally distributed.

J. articulatus, L.
Groton and Lowell (Dr. C. W. Swan); Medford (Wm. Boott).

J. militaris, Bigel. BAYONET RUSH.
Tewksbury (B. D. Greene); Dracut and Westford (Dr. C. W. Swan); Concord (Walter Deane); Bedford (C. W. Jenks).

J. acuminatus, Michx., var. **debilis,** Engelm.
"Issue of Mystic river" (Wm. Boott; specimen in herb. of).

J. acuminatus, Michx., var. **legitimus,** Engelm.
Lowell (Dr. C. W. Swan); Medford and Somerville (C. E. Perkins).

J. nodosus, L.
Medford (Wm. Boott; specimen in herb. of).

J. scirpoides, Lam.
Medford (Wm. Boott; specimen in herb. of).

J. scirpoides, Lam., var. *polycephalus,* Engelm.
Tewksbury (B. D. Greene), specimen in herb. B. S. N. H. Undoubtedly adv. from farther South.

J. Canadensis, J. Gay, var. **longicaudatus,** Engelm.
Generally distributed.

J. Canadensis, J. Gay, var. **coarctatus,** Engelm.
Billerica and Marlboro (Dr. C. W. Swan); Somerville (C. E. Perkins).

PONTEDERIACEÆ. PICKEREL-WEED FAMILY.

PONTEDERIA, L.

P. cordata, L. PICKEREL-WEED.
Common.

COMMELYNACEÆ. SPIDERWORT FAMILY.

TRADESCANTIA, L. SPIDERWORT.

T. Virginica, L.
Ashland (F. S. Collins), growing along the track of the B. & A. R. R., remote from houses. Adv. from farther South.

XYRIDACEÆ. YELLOW-EYED-GRASS FAMILY.

XYRIS, L.

X. flexuosa, Muhl. YELLOW-EYED-GRASS.
Nagog Pond, Acton, Aug., 1882 (W. H. Manning); Ashland (Rev. Thos. Morong).

X. flexuosa, Muhl., var. pusilla, Gray.
Wilmington (C. E. Perkins).
X. Caroliniana, Walt.
Ashland (Rev. Thos. Morong); Ashby (L. L. Dame); Martin's Pond, N. Reading (W. H. Manning). Apparently the commoner species here.

ERIOCAULONACEÆ. PIPEWORT FAMILY.

ERIOCAULON, L.

E. septangulare, With. PIPEWORT.
Rather common.

CYPERACEÆ. SEDGE FAMILY.

CYPERUS, L. GALINGALE.

C. diandrus, Torr.
Common.
C. diandrus, Torr., var. castaneus, Torr.
Somerville (C. E. Perkins).
C. Nuttallii, Eddy, Torr.
Rather common near tide-water.
C. aristatus, Rottb. (C. inflexus, Muhl., Man.)
Winchester (C. E. Perkins); Cambridge, specimen in herb. B. S. N. H.
C. dentatus, Torr.
Common.
C. esculentus, L. (C. phymatodes, Muhl., Man.)
Lowell, Tewksbury and Dracut (Dr. C. W. Swan); Concord (F. S. Collins).
C. strigosus, L.
Common.
C. strigosus, L., var. compositus, Britton.
Malden (F. S. Collins). Possibly introduced.
"Umbel compound; spikelets ⅓-½ in. long, 4-5 flowered." N. L. Britton, Bull. Torr. Bot. Club, Vol. XIII., p. 212.
C. speciosus, Vahl. (C. Michauxianus, Schult., Man.)
Cambridge, (H. A. Young); Apparently scarce in the county.
C. Engelmanni, Steud.
Fresh Pond, Cambridge (Walter Deane); Spy Pond, Arlington (H. A. Young). Not common.

C. filiculmis, Vahl.
Common.

DULICHIUM, Richard.

D. spathaceum, Pers.
Common.

FUIRENA, Rottb.

F. squarrosa, Michx. UMBRELLA GRASS.
Round Pond, Tewksbury (Wm. Boott; specimen in herb. of).

ELEOCHARIS, R. Br. SPIKE-RUSH.

E. Robbinsii, Oakes.
Flat Pond, Groton (Dr. C. W. Swan); Tewksbury and Spot Pond, Stoneham (Wm. Boott); Ashland (Rev. Thos. Morong). Not common.

E. tuberculosa, R. Br.
Tewksbury (B. D. Greene); Wilmington (H. A. Young); Bedford (Dr. C. W. Swan). Rare.

E. obtusa, Schult.
Malden, Winchester, Concord, et al. Not uncommon.

E. Engelmanni, Steud.
Pine Hill, Medford, Aug. 19, 1880 (C. W. Swan).

E. Engelmanni, Steud., var. **detonsa,** Gray.
Winter Pond, Winchester, Aug. 23, 1878 (E. H. Hitchings). At the same locality, Aug., 1884, a proliferous form, in dense heads of aggregated spikes and entirely asexual; associated closely with the variety (C. W. Jenks).

E. olivacea, Torr.
Tewksbury (B. D. Greene); Medford (Wm. Boott); Melrose (H. A. Young); Littleton (L. L. Dame). Not common.

E. palustris, R. Br.
Common.

E. rostellata, Torr.
Medford (Wm. Boott; F. S. Collins).

E. tenuis, Schult.
Common.

E. acicularis, R. Br.
Common.

E. pygmæa, Torr.
Spot Pond, Stoneham (L. H. Bailey, Jr.); Medford (Wm. Boott); Watertown (F. S. Collins). Not common.

SCIRPUS, L. BULRUSH. CLUB-RUSH.

S. planifolius, Muhl.
Malden and Melrose (H. A. Young); Medford (C. E. Perkins); Concord (E. S. Hoar); Wakefield (F. S. Collins). Not common.

S. subterminalis, Torr.
Generally distributed.

S. pungens, Vahl.
Common.

S. Torreyi, Olney.
Groton (C. W. Jenks); Westford (Dr. C. W. Swan); Spot Pond, Stoneham, and Sandy Pond, Littleton (Wm. Boott); Concord (C. E. Perkins).

S. lacustris, L. (S. validus, Vahl., Man.) GREAT BULRUSH.
Generally distributed.

S. debilis, Pursh.
Tewksbury and Lowell (Dr. C. W. Swan); Winchester and Medford (C. E. Perkins); Newton (C. J. Sprague); Concord (E. S. Hoar).

S. supinus, L., var. **Hallii,** Gray.
Winter Pond, Winchester (E. H. Hitchings, 1878; C. W. Jenks and Dr. C. W. Swan, 1884).

S. maritimus, L. SEA CLUB-RUSH.
Common near brackish water.

S. sylvaticus, L.
Winchester (C. E. Perkins); Medford and Woburn (Wm. Boott); Concord (E. S. Hoar).

S. sylvaticus, L., var. **digynus,** Boeckl. (S. microcarpus, Presl, Man.)
Lowell (Dr. C. W. Swan); Medford and Woburn (Wm. Boott); Malden (F. S. Collins).

S. atrovirens, Muhl.
Ashby, Lowell and Dracut (Dr. C. W. Swan); Belmont (C. E. Perkins); Bedford (C. W. Jenks); Malden and Medford (F. S. Collins). Occasional.

S. Eriophorum, Michx. WOOL-GRASS.
Common, and generally distributed; including the forms var. **cyperinus,** Gray; Lowell, Acton and Hopkinton (Dr. C. W. Swan); and var. **laxus,** Gray; Lowell and Chelmsford (Dr. C. W. Swan).

ERIOPHORUM, L. COTTON GRASS.

E. alpinum, L.
Groton (C. W. Jenks); Woburn and Arlington (Wm. Boott); Ashland (Rev. Thos. Morong).

E. vaginatum, L.
Malden (R. Frohock); Melrose (H. A. Young); Concord (E. S. Hoar); Natick (Rev. Thos. Morong).

E. virginicum, L.
Common.

E. polystachyon, L.
Common.

E. gracile, Koch.
Reading and Medford (C. E. Perkins); Bedford (Dr. C. W. Swan); Concord (E. S. Hoar); Natick (Rev. Thos. Morong).

FIMBRISTYLIS, Vahl.

F. autumnalis, Roem. & Schult.
Rather common.

F. capillaris, Gray.
Common.

RHYNCHOSPORA, Vahl. BEAK-RUSH.

R. fusca, Roem. & Schult.
Groton and Westford (Dr. C. W. Swan); Cambridge (Wm. Boott); Bedford (C. W. Jenks); Concord (E. S. Hoar). Not common.

R. alba, Vahl.
Frequent.

R. glomerata, Vahl.
Common.

CLADIUM, P. Browne. TWIG-RUSH.

C. mariscoides, Torr.
Groton, Westford, Winchester, Ashland, et al. Not very common.

SCLERIA, L. NUT-RUSH.

S. triglomerata, Michx.
Concord (E. S. Hoar; specimen in herb. of).

S. reticularis, Michx.
Winter Pond, Winchester (Wm. Boott, C. E. Perkins, et al.)

CAREX, L. SEDGE.

C. folliculata, L.
Common.

C. intumescens, Rudge.
Frequent.

C. lurida, Wahl. (C. lupulina, Muhl., Man.)
Common.

C. lurida, Wahl., var. **polystachya**, Bailey.
A form with long peduncles (2½ in.) and oblong-cylindrical spikes (2¼ × ⅞ in.), Chelmsford (Dr. C. W. Swan). Prof. Bailey states that he has never before seen this variety from so far east.

C. oligosperma, Michx.
Concord, abundant in cold bogs near the river (Walter Deane); Tewksbury (Wm. Boott); Bedford (Dr. C. W. Swan).

C. rostrata, With., var. **utriculata**, Bailey, (C. utriculata, Boott, Man.)
Common.

C. monile, Tuck.
Generally distributed, but not very common. In a singular form found near the Concord river at Bedford (Dr. C. W. Swan), the culm itself forms by continuation the axis of the upper fertile spike, which, in the specimens at hand, is surmounted, or not, by a solitary and sessile staminate spike.

C. Tuckermani, Boott.
Concord (E. S. Hoar, and specimen in Thoreau Herb.)

C. bullata, Schk.
Rather common.

C. tentaculata, Muhl.
Very common.

C. Pseudo-Cyperus, L.
Lowell (Dr. C. W. Swan).

C. Pseudo-Cyperus, L., var. **comosa**, Wm. Boott, (C. comosa, Boott, Man.)
Rather common.

C. scabrata, Schw.
Ashby (Dr. C. W. Swan); Malden (C. E. Perkins); Concord, abundant (Walter Deane); Arlington (Wm. Boott).

C. vestita, Willd.
Frequent. A leafy form with usually three fertile spikes, and lowest bract much exceeding the culm, occurs at Lowell (Dr. C. W. Swan).

C. filiformis, L.
Common.

C. filiformis, L., var. **latifolia**, Boeckl. (C. lanuginosa, Michx., Man.)
Common.

C. HIRTA, L.
Cambridge (L. H. Bailey, Jr.); Melrose (C. E. Perkins); Medford (Wm. Boott); Ashland (Rev. Thos. Morong). Nat. from Eu.

C. trichocarpa, Muhl.
Concord (E. S. Hoar; specimen in herb. of); Ashland (Rev. Thos. Morong).

C. riparia, W. Curtis.
Malden (H. A. Young); Medford (F. S. Collins); Concord Turnpike (Wm. Boott); Cambridge (L. H. Bailey, Jr.)

C. Buxbaumii, Wahl.
Arlington, Chelmsford, Ashland, et al. Not very common.

C. vulgaris, Fries.
Common.

C. aquatilis, Wahl.
Concord (specimen in Thoreau Herb., *fide* Walter Deane).

C. stricta, Lam., (including C. stricta, var. strictior, and C. angustata, Boott, Man.)
Very common.

C. stricta, Lam., var. **decora,** Bailey, (C. aperta, Boott, Man.)
Cambridge (L. H. Bailey, Jr.); Melrose and Ashland (Rev. Thos. Morong); Concord (E. S. Hoar); Belmont and Winchester (Wm. Boott); Hopkinton (F. S. Collins).

C. salina × stricta, Bailey, (C. spiculosa, Fries. ?, Wm. Boott, Bot. Gaz., IX., 88).
Salt marshes, Medford and Arlington (Wm. Boott; Rev. Thos. Morong).

C. lenticularis, Michx.
Lowell, banks of the Merrimack (Dr. C. W. Swan).

C. prasina, Wahl. (C. miliacea, Muhl., Man.)
Medford (C. E. Perkins); Belmont and Arlington (Wm. Boott).

C. salina, Wahl.
Medford (Rev. Thos. Morong; Wm. Boott); Arlington (F. S. Collins).

C. maritima, Mueller.
Medford (Wm. Boott); Cambridge (Man.); Arlington (Edwin Faxon; specimen in herb. of). Rare.

C. crinita, Lam.
Common.

C. crinita, Lam., var. **gynandra,** Schw. & Torr. (C. gynandra, Schw., Man.)
Ashby, Lowell and Framingham (Dr. C. W. Swan); Medford and Woburn (Wm. Boott); Concord (Horace Mann, 1862).

C. limosa, L.
Long Pond, Tewksbury (Wm. Boott, specimen in the Boott Herb.)

C. virescens, Muhl.
Common.

C. arctata, Boott.
Ashby (Dr. C. W. Swan); Medford (C. E. Perkins); Malden (Wm. Boott).

C. debilis, Michx.
Common.

C. gracillima, Schw.
Lowell, Acton, Cambridge, Newton, et al. Not uncommon.

C. flava, L.
Common.

C. Œderi, Retz.
Malden (H. A. Young); Medford (Wm. Boott).

C. fulva, Good.
Tewksbury (B. D. Greene, specimen in herb. B. S. N. H.) Not since reported, and probably adventive.

*C. lævigata, Smith.
Tewksbury (B. D. Greene). Not since reported, and probably adventive.

C. pallescens, L.
Common; including the form with wavy sheaths known as var. **undulata,** Carey, Concord (E. S. Hoar).

C. conoidea, Schk.
Rather common.

C. laxiflora, Lam., var. **patulifolia,** Carey, (var. plantaginea, Boott, Man.)
Malden and Hopkinton (F. S. Collins); Medford (Wm. Boott).

C. laxiflora, L., var. **intermedia,** Boott.
Melrose (Rev. Thos. Morong); Malden and Medford (Wm. Boott); Woburn (C. E. Perkins); Wakefield (F. S. Collins).

C. laxiflora, Lam., var. **striatula,** Carey, (var. blanda, Boott, Man.)
Cambridge (L. H. Bailey, Jr.); Melrose (Rev. Thos. Morong); Winchester (Wm. Boott); Malden and Wakefield (F. S. Collins). An abnormal form with enlarged and inflated pergynia, Wakefield, June 19, 1887 (F. S. Collins).

C. laxiflora, Lam., var. **latifolia,** Boott.
Acton (Walter Deane; specimen in herb. of); Malden (Wm. Boott).
It is doubtful if we have the type, which is thus characterized; "Perigynium elliptic, attenuated at the apex, not prominently nerved; beak not strongly curved; leaves 3 lines or less broad; pistillate spikes an inch or more long, narrow, loosely flowered, cylindrical; staminate spike long peduncled." Bailey, Prel. Synopsis N. A. Carices.

C. retrocurva, Dew.
Melrose and Medford (C. E. Perkins); Cambridge (L. H. Bailey, Jr.); Concord (herb. Thoreau, *fide* Walter Deane); Belmont (Wm. Boott).

C. digitalis, Willd.
Cambridge (L. H. Bailey, Jr.); Belmont (Walter Deane); Malden and Medford (Wm. Boott); Wakefield (F. S. Collins). Not very common.

C. platyphylla, Carey.
Medford (C. E. Perkins); Cambridge (L. H. Bailey, Jr.)

C. PANICEA, L.
Cambridge and Concord (Walter Deane); Malden and Melrose (H. A. Young); Watertown (L. H. Bailey, Jr.); Reading (C. E. Perkins); Newton (Wm. Boott). Nat. from Eu.

C. eburnea, Boott.
Reading (W. H. Manning.) Rare.

C. pedunculata, Muhl.
Woburn (Wm. Boott); Fresh Pond, Cambridge. Specimen in herb. B. S. N. H.

C. Pennsylvanica, Lam.
Very common.

C. varia, Muhl.
Ashby (Dr. C. W. Swan); Melrose and Stoneham (H. A. Young); Medford (Wm. Boott); "var. B. minor, Malden and Spot Pond," (herb. Wm. Boott).

C. Emmonsii, Dew.
Ashby and Waltham (Dr. C. W. Swan); Malden (H. A. Young); Medford (C. E. Perkins); Wilmington and Melrose (F. S. Collins).

C. umbellata, Schk.
Medford and Malden (Wm. Boott).

C. PRÆCOX, Jacq.
Malden (H. A. Young). Nat. from Eu.

C. Willdenovii, Schk.
Melrose (Wm. Boott); Waltham list. Specimen in the Boott Herb.

C. polytrichoides, Muhl.
Generally distributed.

C. stipata, Muhl.
Common.

C. teretiuscula, Good.
Winchester, Tewksbury, and Fresh Pond, Cambridge (Wm. Boott); Concord (E. S. Hoar). Specimen in the Boott Herb.

C. vulpinoidea, Michx.
Common.

C. tenella, Schk.
Billerica (Dr. C. W. Swan); Concord (E. S. Hoar, and herb. Thoreau).

C. rosea, Schk.
Common.

C. rosea, Schk., var. **radiata,** Dew. (including var. minor, Boott, Man). Stoneham (F. S. Collins). Probably not uncommon.

C. rosea, Schk., var. **retroflexa,** Torr. (C. retroflexa, Muhl., Man.)
Malden (H. A. Young); Melrose (Rev. Thos. Morong); Concord (E. S. Hoar); Newton (F. S. Collins); hills about Spot Pond (Wm. Boott). Not common.

C. MURICATA, L.
Somerville (Dr. C. W. Swan); Concord (herb. Thoreau); Medford (F. S. Collins); Arlington (Wm. Boott). Nat. from Eu.

C. sparganioides, Muhl.
Melrose (H. A. Young); Winchester (C. E. Perkins); Concord (E. S. Hoar); Malden (F. S. Collins). Not common.

C. Muhlenbergii, Schk.
Rather common.

C. cephalophora, Muhl.
Not uncommon.

C. exilis, Dew.
Concord (E. S. Hoar); Tewksbury (F. Boott; specimen in herb. B. S. N. H.); Lexington (C. E. Faxon); et al. Not very common.

C. echinata, Murr., var. **microstachys,** Boeckl. (including C. sterilis, Willd., C. stellulata, L., var. angustata, Carey, and var. scirpoides, Carey, Man.)

C. canescens, L.
Common.

C. canescens, L., var. **polystachya,** Boott.
Ashland (Rev. Thos. Morong; specimen in herb. of).

C. canescens, L., var. **vulgaris,** Bailey, (var. alpicola, Bailey, Synopsis, in part.)
Cambridge (W. Deane; specimen in herb. of).
"Differs from the species in its more slender culm and laxer habit, its small spikes, and usually smaller and spreading perigynia. Typical C. canescens is a stout plant, with compact spikes, one-fourth or three-eighths inch long. Both the species and this variety are characterized by a silvery color of the spikes." Bailey, Bot. Gaz., XIII., 86.

C. trisperma, Dew.
Not uncommon.

C. bromoides, Schk.
Medford (C. E. Perkins); Belmont (Wm. Boott).

C. Deweyana, Schw.
Concord (E. S. Hoar; specimen in herb. of).

C. siccata, Dew.
Lowell, Groton, Newton, Arlington, et al. Not uncommon.

C. tribuloides, Wahl. (C. lagopodioides, Schk., Man.)
Common.

C. tribuloides, Wahl., var. **cristata,** Bailey, (C. cristata, Schw., Man.)
Ashland and Natick (Rev. Thos. Morong); Belmont (Wm. Boott). Specimen in the Boott Herb.

C. tribuloides, Wahl., var. **reducta,** Bailey.
Waverly (L. H. Bailey, Jr.; specimen in herb. of).
"Culm slender, especially above, where it surpasses the long-pointed and lax leaves: spikes two to ten, small, nearly globular (usually less than three lines in diameter), all distinct, the lowest separated, mostly bright straw or rust color, the points of the spreading perigynium conspicuous." Bailey, Prel. Syn. N. A. Carices.

C. scoparia, Schk.
Common.

C. adusta, Boott.
Ashland, abundant (Rev. Thos. Morong); Malden, Medford and Newton (Wm. Boott). Specimen in the Boott Herb.

C. straminea, Schk. (including vars. festucacea, Boott, hyalina and typica, Gray, and C. fœnea, var. (?) ferruginea, Gray, Man.)
Common.

C. straminea, Schk., var. **mirabilis,** Tuck. (C. cristata, var. mirabilis, Gray, Man.)
Wilmington (F. S. Collins); Cambridge (L. H. Bailey, Jr.); Medford, common (Wm. Boott); Acton (Walter Deane).

C. straminea, Schk., var. **alata,** Bailey, (C. alata, Torr., Man.)
Chelmsford (Dr. C. W. Swan); Cambridge (Walter Deane); Ashland (Rev. Thos. Morong). Not common.

C. straminea, Schk., var. **fœnea,** Torr. (C. fœnea, Willd., Man.)
Malden, Medford and Stoneham (Wm. Boott).

C. straminea, Schk., var. **moniliformis,** Tuck. (C. fœnea, Willd., var. (?) sabulonum, Gray, Man.)
Malden (F. S. Collins).

C. straminea, Schk., var. **aperta,** Boott.
Cambridge, common (L. H. Bailey, Jr.); Watertown (C. E. Perkins); Arlington (F. S. Collins); Medford (Wm. Boott).

C. straminea, Schk., var. **invisa,** Wm. Boott.
Mystic Pond, Winchester (L. H. Bailey, Jr.); Medford (Wm. Boott); Medford "a form which is almost var. invisa," *fide* Bailey, (F. S. Collins).

C. straminea, Schk., var. ———— ?
"A singular form with which I am not acquainted," L. H. Bailey, Jr. Wilmington, June 15, 1887 (F. S. Collins).
Culms about 2 feet high, rather slender, smooth, roughish at top, considerably exceeding the few (3 to 5) narrowly linear (2″ wide, 6′ to 10′ long), taper-pointed, scabrous-edged leaves; spikes 3–5, nearly globular (3″), the upper contiguous or distinct, the lowest sometimes rather remote, the uppermost (5″ long) abruptly and conspicuously contracted at the long, staminate base, the others little or not at all contracted; perigynia spreading, from lanceolate to rather broadly ovate, roundish or somewhat pointed at base, moderately wing-margined, distinctly several-nerved on each side, gradually tapering from the middle to the slightly bifid apex of the rough-margined beak, a third longer than the scarious light-brown ovate obtuse or blunt-pointed scale; achenium shining, elliptic-ovate, long-apiculate below the green style, abruptly contracted at the sessile base. Spikes pale-green and brownish. Bracts short and filiform, or absent. Sheaths smooth.

GRAMINEÆ. GRASS FAMILY.

PASPALUM, L.

P. setaceum, Michx.
Common.

PANICUM, L. PANIC GRASS.

P. filiforme, L.
Generally distributed, but not common.

P. GLABRUM, Gaudin.
Not uncommon. Nat. from Eu.

P. SANGUINALE, L. CRAB GRASS. FINGER GRASS.
Common in cultivated and waste ground. Nat. from Eu.

P. agrostoides, Spr.
Frequent.

P. proliferum, Lam.
Westford and Bedford (Dr. C. W. Swan); Waltham List; Concord (E. S. Hoar); Malden and Medford (F. S. Collins).

P. capillare, L. OLD-WITCH GRASS.
Common.

P. virgatum, L.
Concord (E. S. Hoar); and common near salt water.

P. latifolium, L.
Not uncommon.

P. clandestinum, L.
Townsend, Lowell, Medford, Concord, et al. Not uncommon.

P. viscidum, Ell.
Arlington, Aug. 31, 1881 (C. E. Perkins).

P. scoparium, Lam. (P. pauciflorum, Ell., Man.)
Chelmsford (Dr. C. W. Swan); Reading (C. E. Perkins); Waltham, abundant (Walter Deane); Wakefield (F. S. Collins).

P. dichotomum, L.
Dr. C. W. Swan, who has made a special study of this perplexing species (or group of species), has kindly contributed the result of his investigations of the county forms.
"The following are the principal forms found within the county limits, some of them being considered distinct species:

a. Low, tufted, spreading, slender, geniculate, branching, hairy, with small, simple little-stalked or partly sheathed panicles, and small diffuse spikelets. (P. dichotomum, L., P. nodiflorum, Lam.) Common.

b. Similar but taller, more upright, less branched; panicles more stalked. Forms of *a.*, and approaching *c.* and *f.* Common.

c. Slender, tufted, light green, growing in woods, with spreading, narrow, tapering leaves, nearly straight culms, simple short-stalked panicles, and small scattered spikelets; nearly smooth, but the lower nodes sometimes annularly bearded. Not uncommon.

d. Densely tufted, hairy, with straight culms, numerous narrow erect leaves, small panicles, with sheathed stalks and small scattered spikelets. Wilmington and Malden (F. S. Collins). Also at Gloucester.

e. Larger, hairy, with wider, more spreading leaves and larger panicles, growing in rich soil. Wilmington (F. S. Collins.)

f. Tall, erect, little tufted, stoutish or sometimes rather slender, smooth, with spreading or ascending leaves, long-stalked panicles numerously flowered, the spikelets larger, about a line in length, obovate, elliptical, and extending well down upon the rather virgate branches. Not infrequent in rather moist soil.

g. A well marked, stoutish, thick-leaved form of *f*, with nodes strongly and conspicuously bearded, but otherwise smooth, has not been reported from the county, but should be looked for. Specimens have been received from Nantucket and New Jersey.

h. A large form of *f*, with largest leaves 4½ inches by 5 lines. In moist soil at Wilmington (Dr. C. W. Swan).

i. A form of *j*, with very narrow, slender-pointed leaves, rather small spikelets, long-stalked panicles and smooth foliage. Bear's Den Road, Middlesex Fells (F. S. Collins).

j. A form received from Nantucket, but not reported in the county, is moderately tufted, slender, straight, with few erect, thick, very rough-edged, short and short pointed leaves, ciliate near the base with few long hairs, the sheaths pubescent on the margins, the panicles small and long stalked, the spikelets small, obovate and scattered, of stoutish aspect from the wide sheaths and rigid leaves. It is entered here that it may be looked for. Possibly a different species.

P. dichotomum having never been thoroughly worked up, provisional names are omitted, as possibly misleading attempts at adaptations to forms described merely as presented by the county collections."

P. depauperatum, Muhl.
Malden, Winchester, Waltham, Concord, et al. Not very common.

P. CRUS-GALLI, L. BARNYARD GRASS.
Very common. Nat. from Eu.

P. CRUS-GALLI, L., var. HISPIDUM, Gray.
Waltham List; Medford (F. S. Collins). Scarce. Possibly native.

P. miliaceum, L. BIRD MILLET.
Lowell, "dumps" (Dr. C. W. Swan); Malden, Cambridge and Medford (F. S. Collins). Adv. from Eu.

A coarse, stout grass, 1-3 feet high, with hirsute leaves and sheaths, swollen nodes, large, open, nodding panicles, large, ovate, solitary spikelets, becoming yellow in ripening; glumes pointed, subequal.

SETARIA, Beauv.

S. verticillata, Beauv.
Cambridge, growing wild in a garden, 1883 (Walter Deane; specimen in herb. of); Waltham List. Adv. from Eu.

S. GLAUCA, Beauv. FOXTAIL. PIGEON GRASS.
Not uncommon. Nat. from Eu.

S. VIRIDIS, Beauv. GREEN FOXTAIL. GREEN PIGEON GRASS. WILD TIMOTHY. BOTTLE GRASS.
Frequent in waste places. Nat. from Eu.

S. ITALICA, Kunth. HUNGARIAN GRASS. GERMAN MILLET. BENGAL GRASS.
Lowell, Westford, Medford, et al. Forms widely differing in size and appearance are referred by the best authorities to this species. Nat. from Eu.

CENCHRUS, L.

C. tribuloides, L. HEDGEHOG GRASS. BUR GRASS.
Dracut and Chelmsford, in the vicinity of woollen mills (Dr. C. W. Swan). Possibly introduced here, though it is native in this region.

SPARTINA, Schreb.

S. cynosuroides, Willd. FRESH-WATER CORD GRASS.
Lowell (Dr. C. W. Swan); Concord (E. S. Hoar); and common near salt water.

S. juncea, Willd. RUSH SALT GRASS.
Common near salt water.

S. stricta, Roth, var. **glabra,** Gray. SALT MARSH GRASS.
Common near salt water.

S. stricta, Roth, var. **alterniflora,** Gray.
Medford (Miss A. M. Symmes); Cambridge (C. E. Perkins).

ZIZANIA, L.

Z. aquatica, L. INDIAN RICE. WATER OATS.
Not uncommon.

LEERSIA, Sol.

L. Virginica, Willd. WHITE GRASS.
Tewksbury and Lowell (Dr. C. W. Swan); Belmont and Medford (C. E. Perkins); Waltham List. Damp woods; not common.

L. oryzoides, Sw. RICE CUT GRASS.
Common.

ANDROPOGON, L. BEARD GRASS.

A. provincialis, Lam. (A. furcatus, Muhl., Man.)
Not uncommon.

A. scoparius, Michx.
Common in dry places.

*****A. dissitiflorus,** Michx. (A. Virginicus, L., Man.)
Waltham List; also reported in the county, but without precise location, by H. A. Young. Rare.

CHRYSOPOGON, Trin.

C. nutans, Benth. (Sorghum nutans, Gray, Man.) INDIAN GRASS. WOOD GRASS.
Occasional.

PHALARIS, L.

P. Canariensis, L. CANARY GRASS.
Not uncommon in waste places. It yields the Canary seed, and its constant recurrence in waste heaps is thus accounted for. Adv. from Eu.

P. arundinacea, L. REED CANARY GRASS.
Lowell, Concord, Malden, Natick, et al. Not uncommon.
The form known as var. **picta,** Gray, the RIBBON GRASS of the gardens, occurs at Medford (C. E. Perkins); and elsewhere. It is abundantly naturalized in the Medford locality, but shows a tendency to revert to the type, of which it is probably a form rather than a variety.

ANTHOXANTHUM, L.

A. odoratum, L. SWEET VERNAL GRASS.
Rather common. Nat. from Eu.

HIEROCHLOA, Gmel.

H. borealis, Roem. & Schult. VANILLA GRASS. SENECA GRASS.
Malden (R. Frohock); Medford (C. E. Perkins); Concord (E. S. Hoar). Not common. The long summer leaves of this early flowering, fragrant grass, are used in small basket work.

ALOPECURUS, L.

A. pratensis, L. MEADOW FOXTAIL.
Common. Nat. from Eu.

A. geniculatus, L. FLOATING FOXTAIL.
Somerville, Waltham, Concord, et al. Not common. Nat. from Eu.

A. geniculatus, L., var. **aristulatus,** Munro. (A. aristulatus, Michx., Man.) WILD FOXTAIL.
Medford (R. Frohock); Fresh Pond, Cambridge (Dr. C. W. Swan).

A. agrestis, L.
Lowell, "dumps" (Dr. C. W. Swan). Adv. from Eu.
"Culms erect, roughish above; panicle spike-like, cylindrical, narrowed at each end, its branches bearing 1-2 spikelets; empty glumes adnate the lower half, the slightly winged keel short ciliate. Annual." Koch., Taschenb. d. Deutsch. & Schw. Fl. Awn arising near the base of the flowering glume and twice its length.

ARISTIDA, L.

A. dichotoma, Michx. POVERTY GRASS.
Malden, Winchester, Newton, et al. Not uncommon.

A. gracilis, Ell.
Lowell and Winchester (Dr. C. W. Swan); Waltham List; Concord (E. S. Hoar); Medford (Wm. Boott).

A. purpurascens, Poir.
Medford, Concord, Framingham, Waltham, et al. Not common.

STIPA, L.

S. avenacea, L. BLACK OAT GRASS.
Medford (Wm. Boott); Malden (R. Frohock); Wakefield (F. S. Collins). Scarce.

ORYZOPSIS, Michx. MOUNTAIN RICE.

O. melanocarpa, Muhl.
Melrose (Wm. Boott; specimen in herb. of). Rare.

O. asperifolia, Michx.
Melrose (Dr. C. W. Swan); Winchester (Wm. Boott); Wilmington (F. S. Collins). Scarce.

O. Canadensis, Torr.
Chelmsford, Woburn, Concord, Sudbury, et al. Not common.

MUHLENBERGIA, Schreb. DROP-SEED GRASS.

M. sobolifera, Trin.
Medford and Winchester (Wm. Boott; specimen in herb. of); Waltham (B. D. Greene). Not common.

M. glomerata, Trin.
Woburn and Bedford (Dr. C. W. Swan); Concord (E. S. Hoar); Tewksbury and Belmont (Wm. Boott). Not common.

M. Mexicana, Trin.
Rather common.

M. sylvatica, Torr. & Gray.
Malden, Arlington, Tewksbury, Dracut, et al.

M. Willdenovii, Trin.
Lowell, Tewksbury, Dracut (Dr. C. W. Swan); Melrose, rare (H. A. Young); Waltham List; Newton (Wm. Boott).

BRACHYELYTRUM, Beauv.

B. aristatum, Beauv.
Ashby, Lowell, Framingham, Concord, et al. Not uncommon.

PHLEUM, L.

P. PRATENSE, L. TIMOTHY. HERDS' GRASS.
Common. Nat. from Eu.

SPOROBOLUS, R. Br.

S. cryptandrus, Gray.
Lowell and Dracut (Dr. C. W. Swan); Winchester (C. E. Perkins). Scarce.

S. serotinus, Gray.
Common in low meadows.

S. asper, Kunth. (Vilfa aspera, Beauv., Man.)
Somerville (C. E. Perkins); Medford (F. S. Collins).

S. vaginæflorus, Vasey. (Vilfa vaginæflora, Torr., Man.)
Not uncommon.

AGROSTIS, L.

A. perennans, Tuck. THIN GRASS.
Lowell and Hopkinton (Dr. C. W. Swan); Malden (R. Frohock); Winchester (C. E. Perkins); Concord (E. S. Hoar).

A. scabra, Willd. HAIR GRASS.
Common.

A. canina, L. BROWN BENT GRASS.
Medford (Miss A. M. Symmes); Woburn (Wm. Boott); Concord (E. S. Hoar). Not common.

A. ALBA, L. WHITE BENT GRASS.
Common. Naturalized here, but native farther north and west.

A. ALBA, L., var. VULGARIS, Thurb. (A. vulgaris, With., Man.) RED-TOP.
Common. Nat. here, like the type.

POLYPOGON, Desf.

P. Monspeliensis, Desf. BEARD GRASS.
Lowell, Billerica and Dracut (Dr. C. W. Swan); N. Chelmsford (Rev. W. P. Alcott); near woollen mills. Adv. from Eu.

CINNA, L.

C. arundinacea, L. WOOD REED GRASS.
Concord, Westford, Arlington, et al. Occasional.

C. pendula, Trin. (C. arundinacea, L., var. pendula, Gray, Man.)
Ashby (Dr. C. W. Swan); Waltham List.

GASTRIDIUM, Beauv.

G. australe, Beauv.
Lowell and Billerica, in wool waste (Dr. C. W. Swan). Native of Eu., but probably adv. here from Cal., where it is naturalized.
"Panicle contracted into a somewhat loose, tapering spike, spikelets 1-flowered. Glumes with an enlarged, ventricose, shining base, very acute above, obscurely keeled, the lower longer. Floret less than one-fourth the length of the lower glume, having a very short callus, which is clothed with minute hairs. Lower palet very thin, truncate and dentate at apex, just below which is a very slender awn, equalling or exceeding the glumes (or sometimes absent); upper palet equalling the lower. Scales two, linear, as long as the ovary." Bot. Cal.

DEYEUXIA, Clarion.

D. Canadensis, Hook. (Calamagrostis Canadensis, Beauv., Man.)
BLUE JOINT GRASS.
Common.

D. Nuttalliana, Vasey. (Calamagrostis Nuttalliana, Steud., Man.)
Frequent.

DESCHAMPSIA, Beauv.

D. flexuosa, Griseb. (Aira flexuosa, L., Man.) HAIR GRASS.
Frequent.

D. cæspitosa, Beauv. (Aira cæspitosa, L., Man.) TUFTED HAIR GRASS.
Lowell and Chelmsford (Dr. C. W. Swan).

HOLCUS, L.

H. LANATUS, L. VELVET GRASS.
Rather common. Nat. from Eu.

AVENA, L.

A. striata, Michx. PURPLE WILD OAT.
Concord (E. S. Hoar; specimen in herb. of).

A. sativa, L. COMMON OAT.
Frequently escaped by roadsides and in waste places. Adv. from Eu.

ARRHENATHERUM, Beauv.

A. AVENACEUM, Beauv. TALL OAT GRASS.
Concord (E. S. Hoar); Waltham List; Medford (L. L. Dame).
Nat. from Eu.

DANTHONIA, DC.

D. spicata, Beauv. WILD OAT GRASS.
Common.

D. compressa, Austin. MOUNTAIN OAT GRASS.
Not uncommon.
"Differs from D. spicata in forming a compact sod, by having more numerous and larger leaves, by a larger, longer and spreading panicle, and by the longer, more slender awn-pointed teeth of the flowering glumes." Vasey, Ag. Grasses of the U. S.

ELEUSINE, Gaertn.

E. Indica, Gaertn. DOG'S TAIL GRASS. WIRE GRASS.
Lowell, "dumps" (Dr. C. W. Swan); Malden, near mill (F. S. Collins). Here probably adventive from the South, where it is naturalized from India.

LEPTOCHLOA, Beauv.

L. mucronata, Kunth.
Lowell "dumps" (Dr. C. W. Swan); Malden, Goulding's Mill (F. S. Collins). Adv. from the West.

DIPLACHNE, Beauv.

D. imbricata, Scribn.
Billerica, near woollen mills (Dr. C. W. Swan). Adv. from the West.
"Habit somewhat that of D. fascicularis, but the spikes are much narrower, the spikelets being smaller, closely appressed and overlapping. In the shape of the lower palet they are very distinct; in place of the acute teeth and manifest awn of the other, the teeth in this are minute and blunt, and the midnerve produced only as a mere point." Bot. Cal.

TRIPLASIS, Beauv.

T. purpurea, Chapm. (Tricuspis purpurea, Gray, Man.) SAND GRASS.
Winchester (C. E. Perkins).

PHRAGMITES, Trin.

P. communis, Trin. REED.
Belmont (C. E. Perkins); Cambridge (Bigelow's Fl. Bost.); Concord (E. S. Hoar). Not common.

EATONIA, Raf.

E. obtusata, Gray.
Concord (E. S. Hoar; specimen in herb. of); Watertown (Bigelow's Fl. Bost., under Aira truncata, Muhl.)

E. Pennsylvanica, Gray.
Medford and Malden (Wm. Boott; specimen in herb. of); Concord (E. S. Hoar). Not common.

ERAGROSTIS, Beauv.

E. major, Host. (E. poæoides, Beauv., var. megastachya, Gray, Man.)
Lowell, "dumps" (Dr. C. W. Swan); Malden (H. A. Young); Cambridge (F. S. Collins). Nat. from Eu. in some parts of the U. S., but adv. in the county.

E. minor, Host. (E. poæoides, Beauv., Man.)
Lowell, "dumps;" Westford, woollen mill yard (Dr. C. W. Swan). Introduced with the preceding.

E. pilosa, Beauv.
Lowell, "dump" (Dr. C. W. Swan); Medford (Wm. Boott). Adv. from Eu.

E. Purshii, Schrad.
Lowell, near cotton waste (Dr. C. W. Swan). Adv. from the South.

E. capillaris, Nees.
Not uncommon.

E. pectinacea, Gray.
Reading (R. Frohock); Medford (Miss A. M. Symmes); Waltham List; Concord (E. S. Hoar).

E. pectinacea, Gray, var. **spectabilis,** Gray.
Not uncommon.

DISTICHLYS, Raf.

D. maritima, Raf. (Brizopyrum spicatum, Hook., Man.) SPIKE GRASS. Salt marshes; common.

DACTYLIS, L.

D. GLOMERATA, L. ORCHARD GRASS.
Common, cultivated and spontaneous. Nat. from Eu. A variety with downy spikelets at Medford (Wm. Boott).

BRIZA, L.

B. MEDIA, L. QUAKING GRASS.
Concord (E. S. Hoar; specimen in herb. of); Malden, Melrose and Medford (H. A. Young). Scarce. Nat. from Eu.

POA, L.

P. annua, L. LOW SPEAR GRASS.
Common.

P. compressa, L. WIRE GRASS.
Common.

P. serotina, Ehrh. FALSE RED-TOP. FOWL MEADOW GRASS.
Common.

P. pratensis, L. MEADOW GRASS. KENTUCKY BLUE GRASS.
Very common.

P. TRIVIALIS, L. ROUGHISH MEADOW GRASS.
Medford and Tewksbury (Wm. Boott); Concord (E. S. Hoar). Specimen in the Boott Herb. Nat. from Eu.

P. alsodes, Gray.
Cambridge (Walter Deane); Malden and Newton (Wm. Boott); Medford (C. E. Perkins). Not common.

GLYCERIA, R. Br.

G. Canadensis, Trin. RATTLESNAKE GRASS.
Common.

G. obtusa, Trin.
Common.

G. nervata, Trin. FOWL MEADOW GRASS (in part).
Common.

G. pallida, Trin.
Tewksbury, Waltham, Stoneham, et al. Not very common.

G. arundinacea, Kunth. (G. aquatica, Smith, Man.) REED MEADOW GRASS.
Somerville (C. E. Perkins); Medford (Miss A. M. Symmes); Waltham List.
G. fluitans, R. Br.
Malden, Cambridge, Concord, et al. Not uncommon.
G. acutiflora, Torr.
Medford (Wm. Boott); Malden (C. E. Perkins); Cambridge (Dr. C. W. Swan); Waltham List; Concord (E. S. Hoar).
G. maritima, Wahl. SEA SPEAR GRASS.
Cambridge (Bigelow's Fl. Bost.); Malden (H. A. Young); Somerville (C. E. Perkins).
G. distans, Wahl.
Somerville (C. E. Perkins); Malden (Wm. Boott).

FESTUCA, L.

F. *Myurus*, L.
N. Chelmsford, wool waste (Rev. W. P. Alcott); Billerica, wool waste (Dr. C. W. Swan). Adv. from Eu.
F. tenella, Willd.
Medford, Chelmsford, Winchester, et al. Not common.
F. OVINA, L. SHEEP'S FESCUE.
Medford (C. E. Perkins).
F. OVINA, L., var. GLAUCA.
Reading, introduced and becoming common (W. H. Manning). Nat. from Eu. Leaves thicker and bluish-green.
F. DURIUSCULA, L. (F. ovina, L., var. duriuscula, Gray, Man.)
Not uncommon.
F. RUBRA, L. ? (F. ovina, L., var rubra, Gray, Man.)
There are in the county certain forms with short running rootstocks which ought probably to be referred to this species.
F. *varia*, Haenk., var. *flavescens*. (F. flavescens, Bellard).
Chelmsford, July 3, 1883 (Dr. C. W. Swan). Adv. from Eu.
"Panicle narrow, contracted, except during anthesis; its branches single or in pairs; spikelets 5-8 flowered; flowering glume obscurely 5-nerved, gradually tapering above the middle, short-awned or awnless; ovary bearded at the top; leaves involute, thread-form, roundish; ligule oblong; roots fibrous, without runners. Leaves grass- or bluish-green. Spikelets variegated with white, green and purple; ligules either obtuse or acute; var. *flavescens*: spikelets pale and much less variegated." Koch, Taschenb. d. Deutsch. & Schw. Fl. Might be taken for a form of F. rubra, L.
F. ELATIOR, L.
Reading (W. H. Manning); et al.

F. elatior, L., var. genuina, Hack.
Malden (F. S. Collins).
F. elatior. L., subsp. pratensis, Hack. (F. pratensis, Huds.)
Malden, Medford, et al.
These, in one form or another, are common, introduced and cultivated grasses. Nat. from Eu.
F. nutans, Willd.
Melrose (H. A. Young). Rare.

Bromus. L.

B. secalinus, L. Cheat. Chess.
Not uncommon. Nat. from Eu.
B. *racemosus*, L. Upright Chess.
Melrose (R. Frohock); Malden (F. S. Collins); Cambridge (Walter Deane). In cultivated grounds, etc. Adv. from Eu.
B. *mollis*, L. Soft Chess.
Billerica (Dr. C. W. Swan); Malden (F. S. Collins). Adv. from Eu.
B. Kalmii, Gray. Wild Chess.
Medford (C. E. Perkins); Waltham List; Concord (E. S. Hoar).
B. ciliatus, L.
Frequent.
B. *rubens*, L.
Billerica, in wool waste (Dr. C. W. Swan). Adv. in wool from Cal. A native of Eu.
"Culms densely tufted, 6 to 9 inches high, and, with the narrowly linear leaves and sheaths, pale green and densely soft-pubescent; panicle ovate, 2 to 3 in. long, with very short, erect branches, thickened upwards and rough pubescent; spikelets about 6-flowered, an inch long, including awns. more or less purplish; upper glume one-third the longer, and about one-fourth shorter than its floret; lower palet. like the glumes, rather coarsely pubescent, 7 lines long and 7-nerved, the intermediate nerves less distinct, terminating in two very acute hyaline teeth nearly two lines long, the awn rather longer than the palet; upper palet with long weak hairs; root fibres pubescent." Bot. Cal.
B. tectorum, L.
Rather common, especially near railroads, in the eastern part of the county. Nat. from Eu.
B. *sterilis*, L.
Winchester (C. E. Perkins); Billerica, with wool waste from Cal. (Dr. C. W. Swan); Medford (Miss A. M. Symmes). Adv. from Eu.

B. maximus, Desf.

Medford (C. E. Perkins). Adv. from Eu.

"The spikelets in the three last preceding species are broader upwards and bear long, straight, very conspicuous awns. *Maximus* has a very simple, erect panicle with few large spikelets on short, stoutish, often undivided branches; *sterilis* is similar, but with much smaller spikelets and more slender, longer branches, becoming recurved; *tectorum* has numerous still smaller and proportionately narrower spikelets, on rather short, filiform, drooping and often secund branches, and foliage velvety pubescent." Dr. C. W. Swan.

LOLIUM, L.

L. perenne, L. DARNEL. RAY GRASS. RYE GRASS.

Malden and Melrose (H. A. Young; specimen in herb. of). Occasional in lawns, etc. Adv. from Eu.

L. temulentum, L.

Lowell, "dump" (Dr. C. W. Swan). Adv. from Eu.

AGROPYRUM, Gærtn.

A. repens, Beauv. (Triticum repens, L., Man.) COUCH GRASS. QUITCH GRASS. QUICK GRASS.

Common and variable.

A. repens, Beauv., var. ? (Triticum repens, L., var. nemorale, Anderson, Man.)

Lowell (Dr. C. W. Swan; specimen in herb. of).

A. caninum, Roem. & Schult. (Triticum caninum, L., Man.) AWNED WHEAT GRASS.

Concord (E. S. Hoar; specimen in herb. of).

SECALE, L.

S. cereale, L. RYE.

Lowell, "dumps" (Dr. C. W. Swan); Somerville and Malden (F. S. Collins). Occasionally escaped or spontaneous. Adv. from Eu.

TRITICUM, L.

T. sativum, Lam., (extended). (T. vulgare, Vill., in part). WHEAT.

Occasionally spontaneous by roadsides, etc. The awned form, *T. sativum vulgare* (T. æstivum, L.), the awnless. *T. sativum vulgare muticum*, (T. hibernum, L.) have both been found. Adv. from Eu.

HORDEUM, L.

H. jubatum, L. SQUIRREL-TAIL GRASS.

Malden (H. A. Young); Cambridge (F. S. Collins).

H. maritimum, With.
Billerica, in wool waste (Dr. C. W. Swan). Adv. from Eu.
"Lateral flowers male or neutral; spikelets all awned; glumes rough, the inner ones of the lateral spikelets half-lanceolate and slightly winged, the remainder setaceous. Annual." Koch, Taschenb. d. Deutsch. & Schw. Fl.

H. murinum, L. WAY BENT. WALL BARLEY.
Lowell, Billerica and Dracut, in wool waste (Dr. C. W. Swan); N. Chelmsford, in wool waste (Rev. W. P. Alcott). Adv. from Eu.
"Annual, with stems 1 to 2 feet high, smooth leaves and inflated sheaths: spike 2 to 3 in. long, inclined, compressed, usually included at base by the upper sheath; spikelets, including awns, 2 in. long: glumes of the middle spikelet lanceolate, long-awned, and conspicuously ciliate on the margins; outer glume of the lateral spikelet setaceous, the other similar to those of the central one: lateral florets longer than the central, attenuate into a long awn, scabrous above, and the inner surface covered with long, weak hairs: palet of perfect floret flattened, scabrous above, its awn about three times as long and flattened below." Bot. Cal.

ELYMUS, L. LYME GLASS. WILD RYE.

E. Virginicus, L.
Lowell, Melrose, Watertown, et al. Not uncommon.

E. Canadensis, L.
Not uncommon.

E. Canadensis, L., var. **glaucifolius,** Gray.
Merrimack River, Lowell (Dr. C. W. Swan).

E. striatus, Willd.
Lowell (Dr. C. W. Swan); Melrose (H. A. Young). Not common.

ASPRELLA, Willd.

A. Hystrix, Willd. (Gymnostichum Hystrix, Schreb., Man.) BOTTLE-BRUSH GRASS.
Dunstable, Concord, Melrose, et al. Not very common.

CRYPTOGAMIA.

PTERIDOPHYTES.

EQUISETACEÆ. HORSETAIL FAMILY.

EQUISETUM, L.

E. arvense, L. COMMON HORSETAIL.
Abundant. The var. **serotinum**, Meyer, has been found at Framingham (Miss J. W. Williams).

E. sylvaticum, L.
Frequent.

E. limosum, L.
Not very common.

E. hyemale, L. SCOURING-RUSH. SHAVE-GRASS.
Not very common.

FILICES. FERNS.

POLYPODIUM, L.

P. vulgare, L. POLYPODY.
Common.

PTERIS, L.

P. aquilina, L. BRAKE.
Common.

ADIANTUM, L.

A. pedatum, L. MAIDENHAIR.
Generally distributed, but not abundant.

WOODWARDIA, Smith. CHAIN FERN.

W. angustifolia, Smith.
Medford, one small station (R. Frohock).

W. Virginica, Smith.
Malden, Medford, Chelmsford, Ashby, et al. Not very common.

ASPLENIUM, L. SPLEENWORT.

A. Trichomanes, L.
Once rather common, but becoming rare in the neighborhood of large towns.

A. ebeneum, Ait.
Not uncommon. "Var. **serratum,** Gray, . . . Malden, Nov., 1872." G. E. Davenport, in Catalogue of the Davenport Herbarium.

A. thelypteroides, Michx.
Medford (L. L. Dame); Melrose (F. S. Collins); Dracut (Dr. C. W. Swan); Groton (C. W. Jenks). Not common.

A. Filix-fœmina, Bernh. LADY FERN.
Common and variable; the only form seemingly worthy a special recognition being the following.

A. Filix-fœmina, Bernh., var. **Michauxii,** Mett. (var. augustum, Eaton).
Medford and Sudbury (G. E. Davenport); Malden (F. W. Morandi).

PHEGOPTERIS, Fée. BEECH FERN.

P. polypodioides, Fée.
Malden and Melrose (F. S. Collins); Medford and Ashby (L. L. Dame); Waltham List; Somerville (E. H. Hitchings). Not common.

P. hexagonoptera, Fée.
Medford (G. E. Davenport); Malden and Melrose (F. S. Collins); Groton (C. W. Jenks). Not common.

P. Dryopteris, Fée.
Arlington Heights (Wm. Boott); Melrose (F. W. Morandi). Very rare.

ASPIDIUM. Swartz. SHIELD FERN. WOOD FERN.

A. Noveboracense, Swartz.
Common.

A. Thelypteris, Swartz.
Common.

A. cristatum, Swartz.
Rather common.

A. cristatum, Swartz., var. **Clintonianum,** Eaton.
Malden (F. W. Morandi); Arlington (Mrs. P. D. Richards).

A. marginale, Swartz.
Common.

A. spinulosum, Swartz.
Rather common.

A. spinulosum, Swartz., var. **intermedium,** Eaton.
Ashby (L. L. Dame); Newton and Malden (G. E. Davenport); Billerica (Dr. C. W. Swan). Not quite so common as the type.

A. spinulosum, Swartz., var. **dilatatum,** Horneman.
Groton (C. W. Jenks); Malden and Melrose (F. S. Collins). This variety does not in the county assume its most highly developed form.

A. Boottii, Tuck. (A. spinulosum, var. Boottii, Man.)
Malden, Medford, Lowell, Ashby, et al. Not very common.
A. acrostichoides, Swartz. CHRISTMAS FERN.
Common. The var. **incisum,** Gray, at Medford (Mrs. P. D. Richards).

CYSTOPTERIS, Bernh. BLADDER FERN.

C. fragilis, Bernh.
Not uncommon.

ONOCLEA, L.

O. sensibilis, L. SENSITIVE FERN.
Very common. The form known as var. **obtusilobata** is occasionally found.

O. Struthiopteris, Hoffm. (Struthiopteris Germanica, Willd., Man.) OSTRICH FERN.
Malden (F. W. Morandi), but now probably extinct in this station; Lincoln (G. E. Davenport); Bedford (L. L. Dame). Rare.

WOODSIA, R. Br.

W. Ilvensis, R. Br.
Not uncommon.

W. obtusa, Torr.
Malden and Melrose (F. S. Collins); Medford (L. L. Dame); Chelmsford (Dr. C. W. Swan). Not very common.

DICKSONIA, L'Her.

D. pilosiuscula, Willd. (D. punctilobula, Kunze, Man.)
Common.

LYGODIUM, Swartz.

L. palmatum, Swartz. CLIMBING FERN.
Concord (G. E. Davenport); West Newton (Severance Burrage); Dracut (Dr. C. W. Swan); Groton (C. W. Jenks). Rare.

OSMUNDA, L.

O. regalis, L. FLOWERING FERN. ROYAL FERN.
Common.

O. Claytoniana, L.
Common.

O. cinnamomea, L. CINNAMON FERN.
Common. The state called var. **frondosa** at Newton (F. S. Plympton); Medford (G. E. Davenport).

OPHIOGLOSSUM, L.

O. vulgatum, L. ADDER'S TONGUE.
Malden, Melrose and Concord (F. S. Collins); Newton (F. S. Plympton); Chelmsford (Dr. C. W. Swan). Not common.

BOTRYCHIUM, Swartz. MOONWORT.

B. matricariæfolium, A. Br.
Ayer (W. H. Manning); Stoneham (L. L. Dame); Medford (F. W. Morandi). Rare.

B. lanceolatum, Angstrom.
Stoneham (L. L. Dame); Groton (C. W. Jenks); Sudbury (Minot Pratt). Very rare.

B. ternatum, Swartz. (B. lunarioides, Man.)
Concord (Davenport Catalogue); Newton (F. S. Plympton); Medford (G. E. Davenport). Infrequent.

B. ternatum, Swartz., var. **obliquum,** Milde.
Rather common.

B. ternatum, Swartz., var. **dissectum,** Milde.
Rather common.

B. Virginianum, Swartz.
Not uncommon.

LYCOPODIACEÆ. CLUB-MOSS FAMILY.

LYCOPODIUM, L., Spring.

L. lucidulum, Michx.
Medford (L. L. Dame); Reading (R. Frohock); Tyngsboro' and Groton (Dr. C. W. Swan).

L. inundatum, L.
Medford (C. E. Perkins); No. Reading (W. H. Manning); Tewksbury (B. D. Greene).

L. annotinum, L.
Ashby (Dr. C. W. Swan).

L. dendroideum, Michx. GROUND PINE.
Common. The var. **obscurum,** Gray, at Medford (G. E. Davenport); Sudbury (Bigelow's Fl. Bost.)

L. clavatum, L. CLUB MOSS.
Common.

L. complanatum, L.
Common.

SELAGINELLA, Beauv., Spring.

S. rupestris, Spring.
Stoneham (F. S. Collins); Malden (R. Frohock); Groton (C. W. Jenks). Probably not uncommon.

S. apus, Spring.
Bedford (C. W. Jenks).

ISOETEÆ. QUILLWORT FAMILY.

ISOETES, L. QUILLWORT.

I. lacustris, L.
Fresh Pond, Cambridge (Wm. Boott). An unusually southern station for this species.

I. Tuckermani, A. Br.
Mystic River and Pond (E. Tuckerman), the original station; Horn Pond, Woburn (Wm. Boott); Spy Pond, Arlington (Rev. Thos. Morong).

I. echinospora, Durieu, var. Braunii, Engelm.
Not uncommon.

I. echinospora, Durieu, var. Boottii, Engelm.
Round Pond, Woburn, the original station; Tofit swamp, Lexington (Wm. Boott).

I. echinospora, Durieu, var. muricata, Engelm.
Woburn Creek and Abajona River, the original locality, not uncommon (Wm. Boott); Bedford (C. W. Jenks); So. Natick (Rev. Thos. Morong); Lexington (Wm. Boott).

I. riparia, Engelm.
So. Natick (Rev. Thos. Morong).

I. Engelmanni, A. Br.
Arlington Brook; Alewife Brook, Woburn (Wm. Boott).

MARSILIACEÆ. WATER-FERN FAMILY.

MARSILIA, L.

M. QUADRIFOLIA, L.
Concord, Bedford, Belmont, Cambridge, et al. Int. and nat. from farther south.

BRYOPHYTES.

MUSCI. MOSSES.

The following list is founded principally on the collections of Mr. H. A. Young, Mrs. S. E. French, Miss C. E. Cummings and Messrs. Edwin and Charles E. Faxon, with additional species from the published lists of Mr. T. P. James and Rev. J. L. Russell.

Besides furnishing species not otherwise represented, and new localities for other species, Miss Cummings has revised the lists of Musci and Hepaticæ, which otherwise would have been much more incomplete.

EPHEMERUM, Hampe.

E. serratum, Hampe, var. **angustifolium,** Schimp.
Cambridge (T. P. James; Lesq. & James, Man.)

PHYSCOMITRELLA, Schimp.

P. patens, Schimp.
Chelmsford (Rev. J. L. Russell).

BRUCHIA, Schwaegr.

B. flexuosa, Muell.?
Cambridge (T. P. James); if this is the plant mentioned in the Trans. Am. Phil. Soc., Vol. XIII., under the name of B. flexuosa, Schwaegr., var. nigrescens.

WEISIA, Hedw.

W. viridula, Brid.
Very common.

W. viridula, Brid., var. **gymnostomoides,** Muell.
Cambridge (T. P. James) in Trans. Am. Phil. Soc., Vol. XIII., as Hymenostomum microstomum. R. Br.

DICRANELLA, Schimp.

D. varia, Schimp. ?
Chelmsford (Rev. J. L. Russell); as Dicranum varium, Hedw. Possibly there may have been a mistake in this determination.

D. heteromalla, Schimp.
Waltham (Mrs. S. E. French).

DICRANUM, Hedw.

D. viride, Schimp.
Newton (C. E. Faxon).

D. flagellare, Hedw.
Waltham (Mrs. S. E. French).

D. fulvum, Hook.
Waltham (Mrs. S. E. French).

D. longifolium, Hedw.
Prospect Hill, Waltham (Rev. J. L. Russell).

D. scoparium, Hedw.
Waltham (Mrs. S. E. French); Natick, common (Miss C. E. Cummings).

D. scoparium, Hedw., var. **pallidum,** Lesq. & James.
Waltham (Mrs. S. E. French).
D. spurium, Hedw.
Newton (C. E. Faxon).
D. Drummondii, Muell.
Woburn (Edwin Faxon).
D. undulatum, Turn.
Chelmsford (Rev. J. L. Russell); Waltham (Mrs. S. E. French); Woburn (Edwin Faxon).

FISSIDENS, Hedw.

F. exiguus, Sulliv.
Waltham (Mrs. S. E. French).
F. osmundioides, Hedw.
Melrose (H. A. Young.)
F. adiantoides, Hedw.
Chelmsford (Rev. J. L. Russell); Melrose (H. A. Young); Waltham (Mrs. S. E. French).
F. adiantoides, Hedw., var **immarginatus,** Lindb.
Natick (Miss C. E. Cummings).

LEUCOBRYUM, Hampe.

L. vulgare, Hampe.
Common.
L. minus, Sulliv.
Waltham (Mrs. S. E. French).

CERATODON, Brid.

C. purpureus, Brid.
Common.

POTTIA, Ehrh.

P. cavifolia, Ehrh. ?
Chelmsford (Rev. J. L. Russell), under the name of Gymnostomum ovatum, Hedw. There may be a mistake here, as the localities given for this species in Lesq. & James Man. are from the Rocky Mountains westward.
P. truncata, Fuern.
Chelmsford (Rev. J. L. Russell), as Gymnostomum truncatulum, Hampe; also found at the Botanic Garden, Cambridge, perhaps introduced (A. B. Seymour).

LEPTOTRICHUM, Hampe.

L. tortile, Muell.
Common.

L. vaginans, Lesq. & James.
Waltham and Maynard (Mrs. S. E. French).
L. pallidum, Hampe.
Common.

BARBULA, Hedw.

B. unguiculata, Hedw.
Waltham (Mrs. S. E. French).
B. cæspitosa, Schwaegr.
Waltham (Mrs. S. E. French); Hopkinton (H. A. Young).

GRIMMIA, Ehrh.

G. conferta, Funck.
Waltham (Mrs. S. E. French).
G. apocarpa, Hedw.
Malden (H. A. Young); Waltham (Mrs. S. E. French).
G. Olneyi, Sulliv.
Lexington (Edwin Faxon).
G. Pennsylvanica, Schwaegr.
Waltham (Mrs. S. E. French).

RACOMITRIUM, Brid.

R. Sudeticum, Bruch & Schimp.
Malden and Newton (C. E. Faxon).

HEDWIGIA, Ehrh.

H. ciliata, Ehrh.
Malden, Waltham, Natick, et al. Common.

COSCINODON, Spreng.

C. pulvinatus, Spreng.?
Rev. J. L. Russell reports Fissidens pulvinatus, Hedw., which is a synonym of this species, as abundant on rocks at Chelmsford; but there is no other record of its occurence in our limits, where it would not be expected from its general range.

DRUMMONDIA, Hook.

D. clavellata, Hook.
Waltham, on trees (Rev. J. L. Russell).

ULOTA, Mohr.

U. crispula, Brid.
Natick (Miss C. E. Cummings).
U. Hutchinsiæ, Schimp.
Generally distributed.

ORTHOTRICHUM, Hedw.

O. sordidum, Sulliv. & Lesq.
Natick (Miss C. E. Cummings).

O. strangulatum, Beauv.
Malden (H. A. Young); Waltham (Mrs. S. E. French); Natick (Miss C. E. Cummings).

O. psilocarpum, James.
Cambridge (T. P. James) in Trans. Am. Phil. Soc., Vol. XIII., as O. pusillum, Bruch & Schimp.

O. leiocarpum, Bruch & Schimp.
Cambridge (T. P. James) in Trans. Am. Phil. Soc., Vol. XIII.

O. obtusifolium, Schrad.
Townsend (Miss H. E. Haynes); Waltham (Mrs. S. E. French).

TETRAPHIS, Hedw.

T. pellucida, Hedw.
Common.

APHANORHEGMA, Sulliv.

A. serratum, Sulliv.
Chelmsford (Rev. J. L. Russell), as Schistidium serratum, Wils.; growing with Phascum patens.

PHYSCOMITRIUM, Brid.

P. pyriforme, Brid.
Malden (H. A. Young).

FUNARIA, Schreb.

F. hygrometrica, Sibth.
Common.

BARTRAMIA, Hedw.

B. pomiformis, Hedw.
Frequent.

PHILONOTIS, Brid.

P. fontana, Brid.
Melrose (H. A. Young); Waltham (Mrs. S. E. French); Natick (Miss C. E. Cummings).

LEPTOBRYUM, Schimp.

L. pyriforme, Schimp.
Waltham (Mrs. S. E. French).

WEBERA, Hedw.

W. nutans, Hedw.
Malden (H. A. Young).

BRYUM, Dill.

B. intermedium, Brid.
Malden (H. A. Young); Waltham (Mrs. S. E. French).

B. bimum, Schreb.
Melrose (H. A. Young); Waltham (Mrs. S. E. French); Natick (Miss C. E. Cummings).

B. argenteum, L.
Same localities as the preceding.

B. cæspiticium, L.
Common.

B. capillare, L.
Prospect Hill, Waltham (Rev. J. L. Russell).

B. pseudotriquetrum, Schwaegr.
Cascade Rocks, Melrose (H. A. Young).

B. roseum, Schreb.
Tewksbury, Malden, Natick, et al.

MNIUM, L.

M. cuspidatum, Hedw.
Common.

M. affine, Bland.
Common.

M. hornum, L.
Tewksbury (B. D. Greene); Malden (H. A. Young). Rare.

M. serratum, Laich.
Prospect Hill, Waltham (Rev. J. L. Russell), as Bryum marginatum, Dick.

M. stellare, Reichard.
Waltham (Mrs. S. E. French).

M. cinclidioides, Hueben.
Malden (H. A. Young).

AULACOMNION, Schwaegr.

A. androgynum, Schwaegr.
Chelmsford (Rev. J. L. Russell); Newton (C. E. Faxon).

A. palustre, Schwaegr.
Common.

A. heterostichum, Bruch & Schimp.
Melrose (H. A. Young); Waltham (Mrs. S. E. French); Natick (Miss C. E. Cummings).

ATRICHUM, Beauv.

A. undulatum, Beauv.
Malden, Melrose, Chelmsford, et al.

A. angustatum, Bruch & Schimp.
Common.

POGONATUM, Beauv.

P. brevicaule, Beauv.
Chelmsford (Rev. J. L. Russell), as Polytrichum Pennsylvanicum, Hedw.

POLYTRICHUM, L.

P. formosum, Hedw.
Malden (H. A. Young); Waltham (Mrs. S. E. French).

P. piliferum, Schreb.
Common.

P. juniperinum, Willd.
Common.

P. commune, L.
Common.

DIPHYSCIUM, Mohr.

D. foliosum, Mohr.
Common.

BUXBAUMIA, Hall.

B. aphylla, L.
Cascade rocks, Melrose (H. A. Young); Waltham (Mrs. S. E. French).

FONTINALIS, Dill.

F. antipyretica, L., var. **gigantea,** Sulliv.
Common.

DICHELYMA, Myrin.

D. capillaceum, Bruch & Schimp.
No. Billerica (H. A. Young); Cambridge (T. P. James); Waltham (Mrs. S. E. French); Natick (Miss C. E. Cummings).

D. pallescens, Bruch & Schimp.
Natick (Miss C. E. Cummings).

D. subulatum, Myrin.?
Rev. J. L. Russell reports Fontinalis subulata, Beauv., which is a synonym of this species, as occurring at Chelmsford, "hanging on bushes, principally Cephalanthus occidentalis, L., in partially dessicated mill-ponds, and full of fruit in Nov." It is probable that there is some mistake in the determination.

LEPTODON, Mohr.

L. trichomitrion, Mohr.
Waltham (Mrs. S. E. French).

NECKERA, Hedw.

N. pennata, Hedw.
Melrose (H. A. Young); Waltham (Mrs. S. E. French).

N. complanata, Hueben.
Waltham (Mrs. S. E. French).

LEUCODON, Schwaegr.

L. brachypus, Brid.
Tewksbury (B. D. Greene).

PTERIGYNANDRUM, Hedw.

P. filiforme, Hedw.
Newton (C. E. Faxon).

THELIA, Sulliv.

T. hirtella, Sulliv.
Common.

T. asprella, Sulliv.
Common.

ANOMODON, Hook. & Tayl.

A. rostratus, Schimp.
Wilmington (H. A. Young); Natick (Miss C. E. Cummings).

A. attenuatus, Hueben.
Chelmsford (Rev. J. L. Russell); Waltham (Mrs. S. E. French).

A. obtusifolius, Bruch & Schimp.
Waltham (Mrs. S. E. French).

PYLAISIA, Bruch & Schimp.

P. intricata, Bruch & Schimp.
Chelmsford (Rev. J. L. Russell); Waltham (Mrs. S. E. French); Hopkinton (H. A. Young).

CYLINDROTHECIUM, Bruch & Schimp.

C. cladorrhizans, Schimp.
Chelmsford (Rev. J. L. Russell); Malden and N. Billerica (H. A. Young); Waltham (Mrs. S. E. French); Newton (C. E. Faxon).

C. seductrix, Sulliv.
Chelmsford (Rev. J. L. Russell); Waltham (Mrs. S. E. French); Hopkinton (H. A. Young).

CLIMACIUM, Web. & Mohr.

C. Americanum, Brid.
Common.

HYPNUM, Dill.

H. scitum, Beauv.
Malden (H. A. Young).

H. delicatulum, L.
Natick (Miss C. E. Cummings). H. tamariscinum, Hedw., reported from Townsend (Miss H. E. Haynes); Malden (H. A. Young); Waltham (Mrs. S. E. French); probably belongs here; also H. proliferum, L., reported from Chelmsford (Rev. J. L. Russell). Lesq. and James, Man. of Mosses of N. A., p. 326, say: "The true H. tamariscinum has not been found in North America, or is here very rare, and the specimens distributed under this name in Sulliv. Musc. Allegh., and Sulliv. & Lesq., Musc. Bor.-Am. Exsicc., and in Austin's Musc. Appal., represent mostly H. delicatulum, while those distributed as H. delicatulum mostly represent H. recognitum."

H. paludosum, Sulliv.
Malden (H. A. Young).

H. lætum, Brid.
Waltham (Mrs. S. E. French).

H. lætum, Brid., var. **dentatum,** Lesq. & James.
Malden (H. A. Young).

H. acuminatum, Beauv.
N. Billerica (H. A. Young).

H. salebrosum, Hoffm.
Common.

H. velutinum, L.
N. Billerica and Hopkinton (H. A. Young).

H. rutabulum, L.
Malden (H. A. Young); Waltham (Mrs. S. E. French).

H. Novæ-Angliæ, Sulliv. & Lesq.
N. Billerica and Wilmington (H. A. Young); Waltham (Mrs. S. E. French).

H. populeum, Hedw.
Malden and Melrose (H. A. Young); Waltham (Mrs. S. E. French).

H. plumosum, Swartz.
Melrose (H. A. Young); Waltham (Mrs. S. E. French).

H. Boscii, Schwaegr.
Malden and Melrose (H. A. Young); Waltham (Mrs. S. E. French).

H. serrulatum, Hedw.
Natick (Miss C. E. Cummings); Malden and Hopkinton (H. A. Young); Waltham (Mrs. S. E. French); Chelmsford (Rev. J. L. Russell).

H. rusciforme, Weis.
Townsend (Miss H. E. Haynes); Melrose (H. A. Young).

H. Alleghaniense, Muell.
Melrose (C. E. Faxon and H. A. Young).

H. denticulatum, L.
Common.

H. sylvaticum, Huds.
  Malden (H. A. Young).
H. Muhlenbeckii, Spruce.
  Malden, common (H. A. Young); Waltham (Mrs. S. E. French).
H. serpens, L.
  Townsend, Chelmsford, Waltham, et al.
H. orthocladon, Beauv.
  Melrose (H. A. Young); Waltham (Mrs. S. E. French); Natick (Miss C. E. Cummings).
H. irriguum, Hook. & Wils.
  Cambridge, T. P. James, in Trans. Am. Phil. Soc., Vol. XIII., as H. irriguum, B. & S.
H. riparium, L.
  Chelmsford (Rev. J. L. Russell); Malden (H. A. Young).
H. hispidulum, Brid.
  Waltham (Mrs. S. E. French).
H. chrysophyllum, Brid.
  Malden (H. A. Young); Waltham (Mrs. S. E. French).
H. aduncum, Hedw.
  Waltham (Mrs. S. E. French).
H. fluitans, L.
  Chelmsford (Rev. J. L. Russell).
H. crista-castrensis, L.
  Waltham (Mrs. S. E. French).
H. molluscum, Hedw.
  Waltham (Mrs. S. E. French).
H. reptile, Michx.
  Malden and N. Billerica (H. A. Young); Waltham (Mrs. S. E. French); Natick (Miss C. E. Cummings).
H. imponens, Hedw.
  N. Billerica (H. A. Young).
H. cupressiforme, L.
  Common.
H. Haldanianum, Grev.
  Common.
H. cordifolium, Hedw.
  Malden (H. A. Young).
H. Schreberi, Willd.
  Frequent.
H. splendens, Hedw.
  Townsend, Natick, Malden, et al. Frequent.
H. triquetrum, L.
  Frequent.

SPHAGNACEÆ. PEAT MOSSES.

SPHAGNUM, Dill.

S. acutifolium, Ehrb.
Waltham (Mrs. S. E. French).
S. acutifolium, Ehrh., var. **purpureum,** Schimp.
Malden (H. A. Young).
S. cuspidatum, Ehrh.
Malden (H. A. Young).
S. squarrosum, Pers.
Tewksbury (B. D. Greene).
S. subsecundum, Nees.
Malden (H. A. Young).
S. subsecundum, Nees., var. **obesum,** Schimp.
Malden (H. A. Young).
S. cymbifolium, Ehrb.
Malden (H. A. Young); Waltham (Mrs. S. E. French).

HEPATICÆ. LIVERWORT FAMILY.

RICCIA, Mich.

R. fluitans, L.
Malden (H. A. Young).

MARCHANTIA, L.

M. polymorpha, L.
Common.

GRIMALDIA, Raddi.

G. barbifrons, Bisch.
Cascade rocks, Melrose (H. A. Young).

FIMBRIARIA, Nees.

F. tenella, Nees.
Cascade rocks, Melrose (H. A. Young).

PELLIA, Raddi.

P. epiphylla, Nees.
Malden (H. A. Young).

STEETZIA, Lehm.

S. Lyellii, Lehm.
Long Pond, Melrose (H. A. Young).

METZGERIA, Raddi.

M. myriopoda, Lindb.
Melrose (H. A. Young).

FRULLANIA, Raddi.

F. Grayana, Mont.
Malden (F. S. Collins).

MADOTHECA, Dumort.

M. platyphylla, Dumort.
Chelmsford (Rev. J. L. Russell); Malden (H. A. Young).

RADULA, Nees.

R. complanata, Dumort.
Chelmsford (Rev. J. L. Russell); Melrose (H. A. Young).

BLEPHAROZIA, Dumort.

B. ciliaris, Dumort.
Malden, et al., common (H. A. Young).

BAZZANIA, B. Gr.

B. trilobata, B. Gr.
Chelmsford (Rev. J. L. Russell); Malden (H. A. Young).

LOPHOCOLEA, Nees.

L. bidentata, Dumort.
Chelmsford (Rev. J. L. Russell).
L. heterophylla, Nees.
Melrose (H. A. Young).

SCAPANIA, Dumort.

S. nemorosa, Nees.
Melrose (H. A. Young).

NARDIA, B. Gr.

N. emarginata, B. Gr.
Melrose (H. A. Young).

THALLOPHYTES.

CHARACEÆ.

NITELLA, Ag.

N. capitata, Ag.
Cambridge (A. Braun, Monograph).

N. flexilis, Ag.
Cambridge (Froebel, in A. Br. Monogr.); Melrose and So. Natick (Rev. Thos. Morong).

N. acuminata, A. Br., subsp. **glomulifera,** A. Br.
Glacialis. Fresh Pond, Cambridge (W. G. Farlow). The specimen in Herb. Decaisne from the Merrimack river, collected by Greene, mentioned by A. Braun in the Monograph, in all probability is from Middlesex county.

N. mucronata, A. Br., subsp. **virgata,** A. Br., var. **tenuior,** A. Br.
Cambridge (A. Br. Monogr.) The var. **robustior, forma longifurca,** A. Br. of this subspecies is mentioned in the Monograph as having been collected in the Merrimack by Greene with the preceding species.

N. gracilis, Ag.
Melrose (F. S. Collins); Spot Pond, Stoneham (Rev. Thos. Morong).

N. tenuissima, Kuetz.
Spot Pond (C. E. Faxon); Mystic Pond, Medford (C. E. Perkins); Ashland and So. Natick (Rev. Thos. Morong).

N. polyglochin, A. Br., subsp. **megacarpa,** A. Br., (N. megacarpa, Allen).
Winchester (Rev. Thos. Morong).

CHARA, Leonh.

C. coronata, A. Br.
Cambridge (B. D. Halstead); Malden (F. S. Collins).

C. coronata, A. Br., var. **Schweinitzii,** A. Br.
Cambridge (Rev. Thos. Morong); E. Medford, abundant in slightly brackish water in clay-pits (F. S. Collins).

C. fragilis, Desv.
Cambridge (C. E. Perkins); Natick (Rev. Thos. Morong).

C. sejuncta, A. Br.
Spot Pond and Ashland (Rev. Thos. Morong).

ALGÆ.

The following list includes both fresh and salt water species; as to the latter, it is believed to be reasonably complete for the very limited "coast line" of Middlesex county. There is no rocky shore or sandy beach, but only a salt marsh with muddy banks, extending two or three miles up the Charles and Mystic rivers, so that only a very limited marine flora is to be expected. Not uncommonly plants of Laminaria, etc., are left by the tide within our limits, but as they are undoubtedly brought from a distance, they are not included here.

As regards fresh water algæ, those species are given which are known to have been found in the county, but they are only a small portion of what would be obtained by a careful observer who could give sufficient time to the study; such genera as Spirogyra, Conferva, Scytonema, and Oedogonium, barely or not at all represented here, are abundantly found in the county; so are also Desmids and Diatoms, which have been entirely omitted here, except that a list of the former has been included, taken from the paper by G. Lagerheim, "Bidrag till Amerika's Desmidié-Flora." The species in question were obtained by Lagerheim from specimens of Utricularia, collected at Tewksbury by B. D. Greene, and preserved in Swedish herbaria.

The fresh and the salt water algæ are not given in separate lists, but the marine and submarine species are designated as such. Where not otherwise stated, the authority for a locality is F. S. Collins; and all species are represented by specimens in the Middlesex Institute herbarium, except those credited to Prof. W. G. Farlow, which are represented by specimens in his herbarium; for Prof. Farlow's kindness in giving access to his very rich collection, and for his assistance in many difficult questions of determination, the authors are greatly indebted.

FLORIDEÆ. RED ALGÆ.

POLYSIPHONIA, Grev.

P. nigrescens, Grev.
Medford, in a ditch in the salt marsh. Rare.

GELIDIUM, Lamour.

G. crinale, Ag.
Medford, on stones, etc., in the Mystic river, between tide marks. Inconspicuous and easily overlooked, but not uncommon.

DELESSERIA, Lamour.

D. sinuosa, Lamour.
Mystic river, below low water mark. Rare.

GRACILARIA, Grev.

G. multipartita, Ag., var. **angustissima,** Harv.
Medford, Everett, et al., in ditches in salt marshes. Not rare.

HILDENBRANTIA, Nardo.

H. rosea, Kuetz.
On pebbles in Mystic river, between tide marks. Not common.

RHODYMENIA, Grev.

R. palmata, Grev. DULSE.
Mystic river, below low water mark. Not common.

CHONDRUS, Stack.

C. crispus, Stack. IRISH MOSS.
Mystic river and salt marshes, common.

CERAMIUM, Lyng.

C. rubrum, Ag.
Mystic river and salt marshes, common.

C. strictum, Harv.
Same locality as the preceding, but less common.

SACHERIA, Sirdt.

S. rigida, Sirdt.?
Melrose, growing in masses on perpendicular rocks in "The Cascade." The determination is not quite certain, as this is one of the many forms which have heretofore been included in Lemanea torulosa, and only recently distinguished by Sirodot.

BATRACHOSPERMUM, Roth.

B. vagum, Ag.
Billerica (Edwin Faxon).

B. moniliforme, Roth.
Not uncommon in running water in spring and early summer. This genus and the preceding are the only fresh water Florideæ yet found in the county.

PORPHYRA, Ag.

P. laciniata, Ag.
Somerville, on woodwork in Mystic river, between tide marks.

OOSPOREÆ.

Fucus, L.

F. vesiculosus, L. BLADDERWRACK.
Very common everywhere in salt water.

Ascophyllum, Stack.

A. nodosum, Stack.
With the preceding, and equally common.

Vaucheria, DC.

V. Dillwynii, Grev.
Newton, fresh water (W. G. Farlow).

V. geminata, DC., var. **racemosa,** Walz.
Malden, Medford, and Melrose, rather common in fresh water.

V. terrestris, Lyng.
Malden, fresh water.

V. uncinata, Kuetz.
Newton, fresh water (W. G. Farlow).

V. Thuretii, Woronin.
Ditches in salt marshes, common.

V. litorea, Nordst.
With the preceding, but not so common, and preferring salt rather than brackish localities. The sexual fruit is rarely found in these two species, and as sterile plants of the different species of Vaucheria are hardly distinguishable from each other, it is not impossible that we have other marine or submarine species.

Bulbochæte, Ag.

B. rectangularis, Wittr.
Medford, fresh water brook.

PHÆOSPOREÆ.

Punctaria, Grev.

P. latifolia, Grev.
Medford, ditches in salt marsh.

P. latifolia, Grev., var. **Zosteræ,** LeJolis.
With the type.

Phyllitis, LeJolis.

P. fascia, Kuetz.
Mystic river and salt marshes.

SCYTOSIPHON, Thuret.

S. lomentarius, Ag.
Mystic river and salt marshes.

RALFSIA, Berk.

R. verrucosa, Aresch.
On pebbles, etc., in Mystic river and salt marshes.

R. clavata, Crouan.
Same localities as the last, and rather more common.

LEATHESIA, S. F. Gray.

L. difformis, Aresch.
Ditches in Mystic river salt marshes; a rather reduced form, and not very common.

ECTOCARPUS, Lyng.

E. littoralis, Lyng.
Mystic and Charles rivers, on stones and woodwork between tide marks. Not uncommon.

E. confervoides, Le Jolis.
Medford salt marshes.

E. confervoides, Le Jolis, var. **siliculosus,** Kjellman.
With the type.

PHÆOSACCION, Farlow.

P. Collinsii, Farlow.
Mystic river, growing on Zostera marina. This species, the only one of the genus, has been found only along the shore from Boston to Nahant.
"Fronds olive-brown, tubular, or saccate, composed of a single layer of cells disposed in fours. Hairs wanting. Reproduction by zoospores produced singly (?) in each cell. Fronds subgelatinous, gregarious, compressed-cylindrical, ¼ inch to 1 inch broad, about 2 to 8 inches long, at first saccate, becoming cylindrical, apex at length ruptured. Cells squarish, .0038 to .007 mm. broad; frond .008 to .01 mm. in thickness." W. G. Farlow in Bull. Torr. Bot. Club, IX., 65.

CHLOROSPOREÆ.

BRYOPSIS, Lamour.

B. plumosa, Ag.
Between tide marks on the muddy banks of Mystic river, not common.

DRAPARNALDIA, Ag.

D. glomerata, Ag.
Everywhere common in fresh water brooks in spring and early summer.

STIGEOCLONIUM, Kuetz.

S. tenue, Kuetz.
In running brooks, fresh water, not uncommon.

CHÆTOPHORA, Schrank.

C. endivæfolia, Ag.
In brooks in early spring; rather common and quite variable. All our species of this genus are fresh water plants.

C. pisiformis, Ag.
Newton (W. G. Farlow); Ashby, Medford, and Melrose. Same localities as the preceding, but less frequent.

C. tuberculosa, Ag.?
Billerica (Edwin Faxon). May possibly be a form of some other species.

CLADOPHORA, Kuetz.

C. ægagropila, Kuetz.
Lake Quannapowitt, Wakefield.

C. lætevirens, Harv.
Medford salt marshes.

C. gracilis, Kuetz.
Medford and Everett, in salt water.

C. fracta, Kuetz.
In slightly brackish water, clay pits, Medford.

C. expansa, Kuetz.
Medford and Everett, common in shallow pools in salt marshes.

CHROOLEPUS, Ag.

C. umbrinum, Kuetz.
Newton (W. G. Farlow).

RHIZOCLONIUM, Kuetz.

R. riparium, Harv.
Salt marshes, common.

R. tortuosum, Kuetz.
Medford salt marshes, not common.

R. Kochianum, Kuetz.
Medford salt marshes, in tide pools, mixed with other algæ.

R. lacustre, Kuetz.
Newton (W. G. Farlow).

ULOTHRIX, Thuret.

U. flacca, Thuret.
Salt marshes, abundant.

U. subtilis, Kuetz.
Newton (W. G. Farlow); and elsewhere.

CONFERVA, Link.

C. floccosa, Ag.
Common in fresh water.

C. globulosa, Kuetz.
Cambridge brick yards (W. G. Farlow).

ULVA, (L.) Le Jolis.

U. Lactuca, (L.) Le Jolis.
Very common in salt and brackish water.

U. enteromorpha, Le Jolis, var. **compresssa,** Le Jolis.

U. enteromorpha, Le Jolis, var. **intestinalis,** Le Jolis.

U. enteromorpha, Le Jolis, var. **lanceolata,** Le Jolis.
The first two varieties very common, and the last not uncommon, in salt and brackish water; all varying much in size and form.

U. clathrata, Ag.
Common in salt water, though not so abundant as the preceding species. Quite variable, including among other forms the var. **Rothiana,** forma **prostrata,** Le Jolis, at Medford.

U. Hopkirkii, Harv.
Medford salt marshes.

MONOSTROMA, Wittr.

M. Grevillei, Le Jolis.
Mystic river, on pebbles, etc., between tide marks, in early spring.

M. Vahlii, Ag.?
Ditch in Mystic river salt marsh, Medford. The determination is not absolute, as Prof. J. G. Agardh, to whom specimens were submitted, says: " Your plant comes very near to, or is quite the same as a plant from Greenland, which I have described as M. Vahlii." It does not appear to have been found since the original discovery, unless this is the same species. In this locality it begins to grow quite early in the spring, the fronds being of considerable size when the ice breaks up, which is usually early in March; and by the middle of April it has entirely disappeared.

"Frond rather large, cylindric-obovate, tubular, noticeably dilated from the slender stipe, entire, or the apex finally ruptured; membrane delicate, collapsing, the cells somewhat clathrate in arrangement, the younger angular, separated by thickish walls, appearing somewhat as if arranged in irregular longitudinal lines, the lower elongate, the upper gradually shorter, when mature roundish and not in lines." J. G. Agardh, Till Algern. Syst. IV., 109.

M. latissimum, Wittr.
Medford, claypits, etc. Common in summer in somewhat brackish water. Not before reported from America.
"Frond delicate, of irregular outline, 1-3 dm. in diameter, much plicate, with entire or undulate margin, lubricous, bright green, in the upper part .02 to .026 mm. thick; cells seen from the surface, without definite arrangement, rather closely set, irregularly 4 or 6 angled, with somewhat rounded corners; in cross section of the frond oval or almost circular, .014 to .018 mm. long." F. Hauck, Meeresalgen Deutschlands.

PROTOCOCCACEÆ.

VOLVOX, Ehren.

V. globator, Ehren.
Malden, pond in Middlesex Fells (R. Frohock).

PEDIASTRUM, Meyer.

P. Ehrenbergii, A. Br.
Holliston, on walls of B. & A. R. R. tunnel.

OPHIOCYTIUM, Naeg.

O. cochleare, A. Br.
Hudson.

PROTOCOCCUS, Ag.

P. viridis, Ag.
Very common everywhere.

TETRASPORA, Ag.

T. bullosa, Ag.
Malden.

T. lubrica, Ag.
Malden, et al.; common in spring.

PORPHYRIDIUM, Naeg.

P. cruentum, Naeg.
Medford.

GLŒOCYSTIS, Naeg.

G. Paroliniana, Naeg.
Newton (W. G. Farlow).

CONJUGATÆ.

SPIROGYRA, Link.

S. insignis, Kuetz.
Malden.

S. insignis, Kuetz., var. **Hantzschii,** Petit.
With the type.

S. Weberi, Kuetz.?
Malden. No mature fruit having been found, this determination is not certain.

S. crassa, Kuetz.
Melrose.

ZYGNEMA, Kuetz.

Z. stellinum, Ag.
Malden and Melrose.

PLEUROCARPUS, A. Br.

P. mirabilis, A. Br.
Wakefield, et al.

DESMIDIEÆ.

The Desmids included in this list were found by Dr. G. Lagerheim attached to specimens of Utricularia, etc., collected at Tewksbury by B. D. Greene. Several of the species have not been found elsewhere.

DESMIDIUM, Ag.

D. aptogonum, Bréb.
D. aptogonum, Bréb., var. **acutius,** Nordst.
D. graciliceps, Lagerheim, forma **major,** Lagerheim.

HYALOTHECA, Ehren.

H. dissiliens, Bréb.
H. mucosa, Ehren.

BAMBUSINA, Kuetz.

B. Brebissonii, Kuetz.

SPONDYLOSIUM, Bréb.

S. nitens, Lund.

SPHÆROZOSMA, Corda.

S. excavatum, Ralfs.

MICRASTERIAS, Ag.

M. truncata, Bréb.
M. truncata, Bréb., var. minor, Wolle.
M. depauperata, Nordst.
M. dichotoma, Wolle.
M. muricata, Ralfs. (Euastrum muricatum, Wolle, Desm.)

EUASTRUM, Ralfs.

E. ventricosum, Lund.
E. pinnatum, Ralfs.
E. inerme, Lund.
E. Wollei, Lagerheim (E. intermedium, Wolle, Desm.), var. quadrigibberum, Lagerheim.
E. compactum, Wolle, var. major, Lagerheim.

COSMARIUM, Ralfs.

C. ornatum, Ralfs.
C. orthostichum, Lund.
C. orthostichum, Lund, var. trigonum, Lagerheim.
C. Lagoense, Nordst.
C. Willei, Lagerheim.
C. Portianum, Arch., var. Brasiliense, Wille.
C. Pardialis, Cohn.
C. quinarium, Lund.
C. quadrifarium, Lund.
C. excavatum, Nordst.
C. subcruciforme, Lagerheim.
C. Wolleanum, Lagerheim (C. pseudogranatum, Wolle, Desm.), subspec. granuliferum, Lagerheim.
C. taxichondrum, Lund.
C. taxichondrum, Lund, var. bidentulum, Lagerheim.
C. pseudotaxichondrum, Nordst., subspec. trichondrum, Lagerheim.
C. pseudotaxichondrum, Nordst., var. quadridentulum, Lagerheim.
C. polymorphum, Nordst.
C. oculiferum, Lagerheim.
C. moniliforme, Ralfs, forma elliptica, Lagerheim.
C. Americanum, Lagerheim.
C. pseudopyramidatum, Lund.
C. Cambricum, Cooke & Wills.
C. octogonum, Delp., var. constrictum, Lagerheim.

ARTHRODESMUS, Arch.

A. incrassatus, Lagerheim.
A. incrassatus, Lagerheim, var. cycladatus, Lagerheim.
A. quadridens, Wood, var. æqualis, Lagerheim.
A. notochondrus, Lagerheim.
A. Incus, Hass.
A. octocornis, Ehren.
A. minutus, Ralfs.

XANTHIDIUM, Ralfs.

X. armatum, Bréb.
X. antilopæum, Kuetz.

STAURASTRUM, Ralfs.

S. Clepsydra, Nordst.
S. striolatum, Arch., forma trigona.
S. aristiferum, Ralfs, forma trigona.
S. trifidum, Nordst., var. glabrum, Lagerheim.
S. luteolum, Lagerheim.
S. echinatum, Bréb.
S. tricorne, Menegh.
S. Cerastes, Lund, forma tetragona.
S. leptacanthum, Nordst., forma 6 + 4—radiata.
S. macrocerum, Wolle.
S. leptocladum, Nordst., var. cornutum, Wille.
S. grallatorium, Nordst., var. forcipigerum, Lagerheim.
S. inconspicuum, Nordst.
S. Brasiliense, Nordst.

TETMEMORUS, Ralfs.

T. granulatus, Ralfs.
T. Brebissonii, Ralfs.

PLEUROTÆNIUM, Naeg.

P. Ehrenbergii, Nordst.
P. Indicum, Lund.
P. constrictum, Lagerheim (Docidium constrictum, Wolle, Desm.)
P. verticillatum, Rab,
P. gracile, Rab.

DOCIDIUM, Lund.

D. dilatatum, Lund.

PENIUM, De Bar.

P. minutum, Cleve.
P. margaritaceum, Bréb., var. punctulatum, Ralfs.
P. Digitus, Bréb.

CLOSTERIUM, Nitzsch.

C. acerosum, Ehren.
C. costatum, Corda.
C. angustatum, Kuetz.
C. juncidum, Ralfs.

NOSTOCACEÆ.

CALOTHRIX, Thuret.

C. confervicola, Ag.
Medford and Everett, on Ulva, etc., in salt water.
C. crustacea, Thuret.
With the preceding.

MASTIGONEMA, Kirchner.

M. ærugineum, Kirchner.
Billerica, on Batrachospermum, (E. Faxon).

RIVULARIA, Roth.

R. radians, Thuret.
Spot Pond, Stoneham.

GLOIOTRICHIA, Ag.

G. pisum, Thuret.
Medford and Newton (W. G. Farlow).

SCYTONEMA, Ag.

S. ambiguum, Kuetz.
Newton (W. G. Farlow).
S. Hoffmanni, Ag.
Newton (W. G. Farlow).
S. Guyanense, Mont.
Newton, in a greenhouse, (W. G. Farlow).

TOLYPOTHRIX, Kuetz.

T. Ægagropila, Kuetz.
Spot Pond, Stoneham.
T. truncicola, Thuret.
Newton (W. G. Farlow).

NOSTOC, Vauch.

N. sphæroides, Kuetz.
Cambridge (W. G. Farlow).

N. sphæricum, Vauch.
Melrose, on rocks near the Cascade.

N. muscorum, Ag.
Newton (W. G. Farlow).

N. collinum, Kuetz.
Malden, swamp in Middlesex Fells; Newton (W. G. Farlow).

ANABÆNA, Bory.

A. Flos-aquæ, Kuetz., var. **circinalis,** Kirchner.
Horn Pond, Woburn (W. G. Farlow).

A. gigantea, Wood.
Framingham (W. G. Farlow).

SPHÆROZYGA, Ag.

S. Carmichaelii, Harv.
Everett, salt marshes; Cambridge (W. G. Farlow).

CYLINDROSPERMUM, Kuetz.

C. majus, Kuetz.
Newton (W. G. Farlow); Holliston, on walls of B. & A. R. R. tunnel, form resembling the C. comatum, Wood.

C. muscicola, Kuetz.
Cambridge (W. G. Farlow).

NODULARIA, Mertens.

N. Harveyana, Thuret.
Cambridge, salt marshes (W. G. Farlow).

LYNGBYA, Ag.

L. ochracea, Thuret.
Cambridge (W. G. Farlow).

L. æstuarii, Liebm.
Salt marshes, common.

L. luteo-fusca, Ag.
Mystic river salt marshes.

L. Wollei, Farlow.
Horn Pond, Woburn; Lake Quannapowitt, Wakefield.

MICROCOLEUS, Desmaz.

M. chthonoplastes, Thuret.
Salt marshes, not uncommon.

M. terrestris, Desmaz.
Newton (W. G. Farlow); Melrose.
M. versicolor, Thuret.
Newton (W. G. Farlow).

OSCILLARIA, Kuetz.

O. subtorulosa, Farlow.
Mystic river salt marshes.
O. chlorina, Kuetz.
Newton (W. G. Farlow).
O. tenuis, Ag.
Malden and Reading.
O. nigra, Vauch.
Newton (W. G. Farlow); Malden.
O. viridis, Vauch.
Newton (W. G. Farlow).
O. subuliformis, Harv.
Charles river salt marshes (W. G. Farlow); Mystic river marshes.
O. Frölichii, Kuetz.
Newton (W. G. Farlow).
O. Frölichii, Kuetz., var. **viridis,** Zeller.
Medford clay-pits.
O. Frölichii, Kuetz., var. **ornata,** Rab.
Charles river, Newton.
O. princeps, Vauch.
Cambridge (W. G. Farlow).

BEGGIATOA, Trevis.

B. alba, Trevis., var. **minima,** Warming.
Mystic river salt marshes.
B. mirabilis, Cohn.
Cambridge, salt marshes (W. G. Farlow).

LEPTOTHRIX, Kuetz.

L. rigidula, Kuetz.
Mystic river salt marshes.
L. subtilissima, Rab.
Watertown (W. G. Farlow).

SPIRULINA, Turpin.

S. tenuissima, Kuetz.
Everett and Medford, rather common in brackish water; Cambridge (W. G. Farlow).

GLŒOTHECE, Naeg.

G. confluens, Naeg.
Newton (W. G. Farlow).

APHANOTHECE, Naeg.

A. prasina, A. Br.
Cambridge (W. G. Farlow).

CŒLOSPHÆRIUM, Naeg.

C. Kuetzingianum, Naeg.
Framingham (W. G. Farlow); Spot Pond, Stoneham.

CLATHROCYSTIS, Henfrey.

C. æruginosa, Henfrey.
Horn Pond, Woburn (W. G. Farlow); Spot Pond, Stoneham.

C. roseo-persicina, Cohn.
Very common on salt marshes.

GLŒOCAPSA, Naeg.

G. crepidinum, Thuret.
Everett and Medford, on woodwork near high-water mark.

CHROOCOCCUS, Naeg.

C. turgidus, Naeg.
Everett and Medford, with the preceding species.

LICHENS.

This list is undoubtedly far from being a complete catalogue of the lichens of Middlesex County; and the number of species would probably be double, if the same study could be given to this order as to the flowering plants or ferns. As it stands, it is a list of species known to occur here; but the absence of any species from this list does not at all imply that it does not grow in the county, or even that it is very rare. In arrangement, nomenclature, etc., Tuckerman's later works have been followed, which give a somewhat different system in the use of names of authorities for genera, species, etc., from that employed in other orders. Unless otherwise stated, species from Cambridge, Watertown, Newton, Medford, Arlington, and Lexington are on the authority of Prof. Tuckerman; from Waltham, Mrs. S. E. French; from Chelmsford, Rev. J. L. Russell; and species from Natick and Sherborn were collected by Miss Clara E. Cummings, to whom the writers are much indebted for a revision of the entire list of lichens, as well as for additional species and localities.

RAMALINA, Ach., De Not.

R. calicaris, (L.) Fr., var. **fastigiata,** Fr.
Waltham, Natick.
R. calicaris, (L.) Fr , var. **farinacea,** Schaer.
Waltham.
R. pollinaria, (Ach.)
Natick.
R. polymorpha, (Ach.)
Chelmsford, Waltham.

CETRARIA (Ach.) Fr., Muell.

C. Islandica, (L.) Ach.
Cambridge, Watertown, Newton, Natick.
C. aleurites, (Ach.) Th. Fr.
Natick.
C. aleurites, (Ach.) Th. Fr., var. **placorodia,** Tuck.
Waltham, Chelmsford.
C. ciliaris, (Ach.)
Cambridge, Waltham, Natick, Sherborn.
C. lacunosa, Ach.
Cambridge, Waltham, Natick, Sherborn.
C. juniperina, (L.) Ach.
Cambridge, Natick.

EVERNIA, Ach., Mann.

E. furfuracea, (L.) Mann.
Chelmsford, Waltham.
E. prunastri, (L.) Ach.
Cambridge, Watertown, Natick.

USNEA, (Dill.) Ach.

U. barbata, (L.) Fr., var. **florida,** Fr.
Waltham, Natick, Sherborn.
U. barbata, (L.) Fr., var. **hirta,** Fr.
Waltham, Natick.
U. barbata, (L.) Fr., var. **dasypoga,** Fr.
Natick.
U. angulata, Ach.
Chelmsford.
U. trichodea, Ach.
Chelmsford.
U. longissima, Ach.
Waltham.

ALECTORIA, (Ach.) Nyl.

A. jubata, (L.), var. **chalybeiformis,** Ach.
Natick.

THELOSCHISTES, Norm.

T. chrysophthalmus, (L.), Norm.
Waltham, Natick.

T. parietinus, (L.) Norm.
Cambridge, Waltham, Natick.

T. polycarpus, (Ehrh.)
Waltham.

T. lychneus, (Nyl.)
Cambridge, Natick.

T. concolor, (Dicks.)
Natick, Sherborn.

PARMELIA, (Ach.) De Not.

P. perlata, (L.) Ach.
Framingham and Chelmsford (Russell): Waltham, Cambridge and Lexington.

P. perforata, (Jacq.) Ach.
Waltham, Sherborn, Natick.

P. crinita, Ach.
Cambridge, Waltham.

P. tiliacea, (Hoffm.) Floerk.
Waltham, Sherborn.

P. Borreri, Turn., var. **rudecta,** Tuck.
Waltham, Sherborn.

P. saxatilis, (L.) Fr.
Waltham, Sherborn, Natick.

P. saxatilis, (L.) Fr., var. **sulcata,** Nyl.
Waltham.

P. physodes, (L.) Ach.
Waltham, Sherborn.

P. colpodes, (Ach.) Nyl.
Waltham.

P. olivacea, (L.) Ach.
Waltham, Natick.

P. caperata, (L.) Ach.
Waltham, Sherborn, Natick.

P. conspersa, (Ehrh.) Ach.
Cambridge, Waltham, Natick.

P. ambigua, (Wulf.) Ach., var. **albescens,** Wahl.
Cambridge.

PHYSCIA, (DC., Fr.) Th. Fr.

P. speciosa, (Wulf., Ach.) Nyl.
Watertown, Medford, Cambridge and Waltham (Tuckerman).
P. hypoleuca, (Muhl.) Tuck.
Natick.
P. comosa, (Eschw.) Nyl.
Cambridge, very rare (Tuckerman).
P. aquila, (Ach.) Nyl., var. **detonsa,** Tuck.
Natick.
P. pulverulenta, (Schreb.) Nyl.
Cambridge, Waltham.
P. stellaris, (L.)
Natick, Sherborn.
P. stellaris, (L.), var. **aipola,** Nyl.
Cambridge, Waltham.
P. tribacia, (Ach.) Tuck.
Waltham, Sherborn.
P. hispida, (Schreb., Fr.) Tuck.
Cambridge, Waltham.
P. obscura, (Ehrh.) Nyl.
Waltham, Sherborn, Natick.
P. setosa, (Ach.) Nyl.
Natick.

PYXINE, Fr., Tuck.

P. sorediata, Fr.
Waltham.

UMBILICARIA, Hoffm.

U. Muhlenbergii, (Ach.) Tuck.
Waltham (C. J. Sprague); Cambridge, Medford.
U. Dillenii, Tuck.
Waltham.
U. pustulata, (L.) Hoffm., var. **papulosa,** Tuck.
Waltham, Natick.

STICTA, (Schreb.) Fr.

S. amplissima, (Scop.) Mass.
Waltham, Natick.
S. pulmonaria, (L.) Ach.
Arlington, Waltham, Natick.
S. fuliginosa, (Dicks.) Ach.
Waltham (Russell).
S. crocata, (L.) Ach.
Waltham, Sherborn.

NEPHROMA, Ach.

N. tomentosum, (Hoffm.) Koerb.
Waltham.

N. Helveticum, Ach.
Chelmsford, Medford, Lexington, Natick.

PELTIGERA, (Willd., Hoffm.) Fée.

P. aphthosa, (L.) Hoffm.
Waltham (C. J. Sprague).

P. polydactyla, (Neck.) Hoffm.
Waltham.

P. rufescens, (Neck.) Hoffm.
Cambridge, Waltham.

P. canina, (L.) Hoffm.
Waltham.

P. canina, (L.) Hoffm., var. **sorediata,** (Schaer.)
Waltham.

PHYSMA, Mass.

P. luridum, (Mont.)
Waltham.

PANNARIA, Delis.

P. lanuginosa, (Ach.) Koerb.
Waltham.

P. molybdea, (Pers.) Tuck., var. **cronia,** Nyl.
Waltham.

COLLEMA, Hoffm., Fr.

C. verruciforme, Nyl.
Cambridge.

C. flaccidum, Ach.
Waltham.

LEPTOGIUM, Fr., Nyl.

L. pulchellum, (Ach.) Nyl.
Waltham.

L. Tremelloides, (L. fil.) Fr.
Medford, Waltham, Natick.

L. myochroum, (Ehrh., Schaer.) Tuck., var. **saturninum,** Schaer.
Chelmsford, Cambridge, Watertown, Waltham (Tuckerman).

PLACODIUM, (DC.) Naeg. & Hepp.

P. elegans, (Link) DC.
Cambridge.

P. murorum, (Hoffm.) DC.
Waltham.
P. cinnabarrinum, (Ach.) Anz.
Newton (T. P. Adams).
P. aurantiacum, (Lightf.) Naeg. & Hepp.
Waltham (C. J. Sprague); Natick, Sherborn.
P. cerinum, (Hedw.) Naeg & Hepp.
Waltham, Natick.
P. ferrugineum, (Huds.) Hepp., var. **Pollinii,** Tuck.
Natick.
P. vitellinum, (Ehrh.) Naeg. & Hepp.
Waltham, Natick.

LECANORA, Ach., Tuck.

L. rubina, (Vill.) Ach.
Cambridge, Watertown.
L. muralis, (Schreb.) Schaer., var. **saxicola,** Schaer.
Cambridge, Waltham.
L. pallida, (Schreb.) Schaer.
Waltham, Natick, Sherborn.
L. subfusca, (L.) Ach.
Waltham, Natick, Sherborn.
L. badia, (Pers.) Ach.
Waltham.
L. varia, (Ehrh.) Nyl.
Waltham, Natick, Sherborn.
L. pallescens, (L.) Schaer.
Sherborn.
L. tartarea, (L.) Ach.
Medford, Natick, Sherborn.
L. verrucosa, (Ach.) Laur.
Waltham.
L. cinerea, (L.) Sommerf.
Cambridge.
L. Bockii, (Fr.) Th. Fr.
Waltham (C. J. Sprague).

RINODINA, Mass., Stizenb., Tuck.

R. oreina, (Ach.) Mass.
Waltham.
R. constans, (Nyl.) Tuck.
Natick.

PERTUSARIA, DC.

P. multipuncta, (Turn.) Nyl.
Waltham.

P. communis, DC.
Natick.
P. pustulata, (Ach.) Nyl.
Sherborn.

CONOTREMA, Tuck.

C. urceolatum, (Ach.) Tuck.
Cambridge, Waltham, Natick, Sherborn.

URCEOLARIA, (Ach.) Flot.

U. scruposa, (L.) Nyl.
Chelmsford, Cambridge, Watertown.

MYRIANGIUM, Mont. & Berk.

M. Duriæi, (Mont. & Berk.) Tuck.
Newton (C. J. Sprague).

STEREOCAULON, Schreb.

S. paschale, (L.) Fr.
Medford, Waltham.

CLADONIA, Hoffm.

C. alcicornis, (Lightf.) Floerk.
Cambridge, Watertown.
C. cariosa, (Ach.) Spreng.
Cambridge.
C. pyxidata, (L.) Fr.
Cambridge, Waltham, Natick, Sherborn.
C. gracilis, (L.) Nyl.
Waltham.
C. gracilis, (L.) Nyl., var. **verticillata,** Fr.
Natick, Waltham.
C. gracilis, (L.) Nyl., var. **hybrida,** Schaer.
Natick.
C. cornuta, (L.) Fr.
Sherborn.
C. cenotea, (Ach.) Schaer.
Waltham.
C. cæspiticia, (Pers.) Fl.
Waltham.
C. furcata, (Huds.) Fr.
Waltham.
C. furcata, (Huds.) Fr., var. **racemosa,** Fl.
Waltham, Sherborn.

C. furcata, (Huds.) Fr., var. **subulata,** Fl.
Waltham.

C. rangiferina, (L.) Hoffm.
Waltham.

C. rangiferina, (L.) Hoffm., var. **sylvatica,** L.
Waltham, Sherborn.

C. rangiferina, (L.) Hoffm., var. **alpestris,** L.
Waltham.

C. uncialis, (L.) Fr.
Cambridge, Waltham.

C. Cornucopioides, (L.) Fr.
Waltham, Natick.

C. deformis, (L.) Hoffm.
Waltham.

C. macilenta, (Ehrh.) Hoffm.
Sherborn.

C. cristatella, Tuck.
Waltham, Natick, Sherborn.

BÆOMYCES, Fée.

B. roseus, Pers.
Waltham, Natick.

BIATORA, Fr.

B. Nylanderi, Anz.
Cambridge.

B. uliginosa, (Schrad.) Fr.
Watertown.

B. denigrata, Fr.
Cambridge.

B. mixta, Fr.
Cambridge.

B. milliaria, Fr.
Cambridge.

B. rubella, (Ehrh.) Rab.
Waltham.

LECIDEA, (Ach., Fr.)

L. sylvicola, Th. Fr.
Waltham (C. J. Sprague; *fide* specimen in herb. B. S. N. H.)

BUELLIA, (De Not.) Tuck.

B. parasema, (Ach.) Koerb.
Waltham, Natick.

B. albo-atra, (Hoffm.)
Arlington.

OPEGRAPHA, (Humb.) Ach., Nyl.

O. varia, (Pers.) Fr.
Sherborn.
O. vulgata, (Ach.) Nyl.
Sherborn.

XYLOGRAPHA, Fr., Nyl.

X. opegraphella, Nyl.
Waltham (C. J. Sprague).

GRAPHIS, Ach., Nyl.

G. scripta, (L.) Ach.
Waltham, Natick.
G. dendritica, Ach.
Sherborn.

ARTHONIA, Ach., Nyl.

A. astroidea, Ach.
Cambridge.
A. punctiformis, Ach.
Sherborn.

ACOLIUM (Fée) De Not.

A. tigillare, (Ach.) De Not.
Cambridge, Waltham, Natick.

CALICIUM. Pers., Ach., Fr.

C. subtile, Fr.
Natick.
C. turbinatum, Pers.
Cambridge.

ENDOCARPON, Hedw., Fr.

E. miniatum, (L.) Schaer.
Arlington, Waltham, Natick.
E. miniatum, (L.) Schaer., var. **complicatum,** Schaer.
Cambridge.
E. miniatum, (L.) Schaer., var. **aquaticum,** Schaer.
Medford, Waltham.

VERRUCARIA, (Pers.) Tuck.

V. nigrescens, Pers.
Cambridge.
V. muralis, Ach.
Watertown.

PYRENULA, (Ach., Naeg. & Hepp.) Tuck.

P. punctiformis, (Ach.) Naeg.
Cambridge.

P. gemmata, (Ach.) Naeg.
Sherborn.

P. thelæna, (Ach.) Tuck.
Natick.

ADDITIONS AND CORRECTIONS.

Page 10. After **Violaceæ, Violet Family,** insert the line
VIOLA, L.

Page 11. After the note on **V. pubescens,** Ait., insert
V. pubescens, Ait., var. **scabriuscula,** Gray.
Waltham (Walter Deane; specimen in herb of). Rare.

Page 26. In the description of *Trigonella Cassia*, Boiss., the last clause should read "traversed by longitudinal anastomosing nerves."

Page 38. After the note on *Œnothera biennis*, L., var. *grandiflora*, Lindl., insert
Œ. Oakesiana, Robbins, (Œ. biennis, L., var. Oakesiana, Man.)
Cambridge (S. Watson).

Page 60. For **C. rapunculoides,** read *C. rapunculoides*.

Page 88. For **R. Brittanica,** read **R. Britannica.**

Page 93. For **C. vulgaris,** Lam., var. AMERICANA, A. DC., read **C. vulgaris,** Lam., var. **Americana,** A. DC.

Page 95. For **S. viminalis,** read *S. viminalis*.

Page 98. For **L. trisculca,** read **L. trisulca.**

Page 111. After the note on **Cyperus dentatus,** insert
C. glaber, L.
Westford, woollen mills, Sept. 15, 1884 (Dr. C. W. Swan). Adv. in wool from Turkey in Asia.
"Low, tufted, smooth; leaves narrow, shorter than the culm; involucre about 3-leaved, conspicuous; umbellets very short-stalked in the sessile umbel, forming a capitate cluster of numerous spikes; these linear (¾ inch long by 1½ lines wide) flattened, serrate, scales ovate, strongly conduplicate, obtuse, mucronate, striate, purplish-brown with broad green keel and narrow pale margin, about half-length imbricate; rachis narrowly winged; stamens 3, style 3-cleft; achenium obovate, triquetrous, apiculate, dull." C. W. Swan.

Page 111. After the note on **C. esculentus,** L., insert
C. esculentus, L., var. *angustispicatus*, Britton.
Lowell, Aug. 23, 1885 (Dr. C. W. Swan). Adv. from the South.
"Spikelets narrowly linear, about one line wide and three-fourths of an inch long; a well marked form." Britton, Bull. Torr. Bot. Club, XIII., 211.

Page 111. After the note on **C. strigosus**, L., insert
  C. strigosus, L., var. **capitatus**, Boeckl.
    Melrose (F. S. Collins).
    "Inflorescence of several capitate clusters, rays short." Britton, Bull. Torr. Bot. Club, XIII., 212.
    A doubtful form which may belong under this variety, or is perhaps a hybrid between C. strigosus and C. filiculmis, was found at Chelmsford, Sept. 20, 1885, by Dr. C. W. Swan, who gives the following description:
    "9 inches high; habit of filiculmis, the slender culms with hard tuberous bases, leaves short and very narrow, umbel with very short rays forming one almost sessile dense head, and involucre that of this species; but the scales longer and narrower, strongly conduplicate, yellowish brown, much exceeding the achenium, which (immature) appears to be of an intermediate character, and is partly enclosed by the membranous wings of the rachis. Spikes 8-10 flowered."

RECAPITULATION.

	Genera.	NATIVE		NATURALIZED		ADVENTIVE		TOTAL.
		Species.	Varieties.	Species.	Varieties.	Species.	Varieties.	
Ranunculaceæ	12	23	2	3		5		33
Berberidaceæ	2			2				2
Nymphæaceæ	4	4	1	1				6
Sarraceniaceæ	1	1						1
Papaveraceæ	3	1		1		1		3
Fumariaceæ	4	2		3				5
Cruciferæ	20	15	2	13		15		45
Violaceæ	1	9	2	2		4		17
Cistaceæ	2	6						6
Droseraceæ	1	2						2
Hypericaceæ	2	6		1				7
Elatinaceæ	1	1						1
Caryophyllaceæ	11	11		10		10		31
Paronychieæ	2	2						2
Ficoideæ	1			1				1
Portulacaceæ	2	1		1		2		4
Malvaceæ	4	1		2		6		9
Tiliaceæ	1	1						1
Linaceæ	1	2				1		3
Geraniaceæ	4	7		1		3		11
Rutaceæ	2	1				1		2
Simarubaceæ	1			1				1
Anacardiaceæ	1	5				1		6
Vitaceæ	2	4						4
Rhamnaceæ	2	1		1				2
Celastraceæ	1	1						1
Sapindaceæ	4	6		1		1		8
Polygalaceæ	1	5						5
Leguminosæ	21	28	2	14		20		64
Rosaceæ	14	42	4	6		5		57
Saxifragaceæ	6	9		1		2		12
Crassulaceæ	3	1		3				4

MIDDLESEX FLORA.

	Genera.	Native.		Naturalized.		Adventive.		Total.
		Species.	Varieties.	Species.	Varieties.	Species.	Varieties.	
Hamamelaceæ	1	1						1
Haloragem	3	5	2	1				8
Onagraceæ	5	12	2			2	2	18
Melastomaceæ	1	1						1
Lythraceæ	3	4				2		6
Cactaceæ	1			1				1
Cucurbitaceæ	2			2				2
Umbelliferæ	17	15		4		2		21
Araliaceæ	1	4				1		5
Cornaceæ	2	8						8
Caprifoliaceæ	6	14				1		15
Rubiaceæ	4	11		1		1		13
Compositæ	57	103	11	22		37	2	175
Lobeliaceæ	1	4				1		5
Campanulaceæ	2	3				1		4
Ericaceæ	18	35	3			1		39
Ilicineæ	2	3						3
Plantaginaceæ	1	2		2		1	1	6
Plumbaginaceæ	1	1						1
Primulaceæ	8	9	2	3		1		15
Lentibulariaceæ	1	8						8
Orobanchaceæ	2	2						2
Scrophulariaceæ	14	22		5		5		32
Verbenaceæ	2	3				2		5
Labiatæ	24	17	1	10		13		41
Borraginaceæ	13	4		6		10		20
Hydrophyllaceæ	2					6		6
Polemoniaceæ	2					3		3
Convolvulaceæ	3	3		1		5		9
Solanaceæ	9	3		6		11		20
Gentianaceæ	5	5				1		6
Apocynaceæ	1	2						2
Asclepiadaceæ	3	8	1	1		1		11
Oleaceæ	3	3		2				5
Aristolochiaceæ	1			1				1
Nyctaginaceæ	1					1		1
Phytolaccaceæ	1	1						1
Chenopodiaceæ	5	7	1	7	2	5		22

	Genera.	Native.		Naturalized.		Adventive.		Total.
		Species.	Varieties.	Species.	Varieties.	Species.	Varieties.	
Amarantaceæ	2	2		5		3		10
Polygonaceæ	4	19		8	1	3		31
Lauraceæ	2	2						2
Thymeleaceæ	2	1		1				2
Santalaceæ	1	1						1
Ceratophyllaceæ	1	1	1					2
Callitrichaceæ	1	2						2
Podostemaceæ	1	1						1
Euphorbiaceæ	2	3		1		2		6
Urticaceæ	10	8		5		1		14
Platanaceæ	1	1						1
Juglandaceæ	2	5						5
Cupuliferæ	6	14						14
Myricaceæ	2	3						3
Betulaceæ	2	7	1					8
Salicaceæ	2	13	2	4		2		21
Coniferæ	9	11				1		12
Araceæ	5	5						5
Lemnaceæ	2	3						3
Typhaceæ	2	6	1					7
Naiadaceæ	7	28	5	1				34
Alismaceæ	3	6	1					7
Hydrocharidaceæ	2	2						2
Orchidaceæ	11	30	1			2		33
Amaryllidaceæ	1	1						1
Hæmodoraceæ	1	1						1
Iridaceæ	2	4				1		5
Dioscoreaceæ	1	1						1
Smilaceæ	1	3	1					4
Liliaceæ	19	19		5		8		32
Juncaceæ	2	16	4				1	21
Pontederiaceæ	1	1						1
Commelynaceæ	1	1				1		1
Xyridaceæ	1	2	1					3
Eriocaulonaceæ	1	1						1
Cyperaceæ	11	109	27	4		3	1	144
Gramineæ	51	86	5	22	5	27		145
Equisetaceæ	1	4						4

MIDDLESEX FLORA.

	Genera.	Native.		Naturalized.		Adventive.		Total.
		Species.	Varieties.	Species.	Varieties.	Species.	Varieties.	
Filices............	15	34	6					40
Lycopodiaceæ........	2	8						8
Isoetæ............	1	5	2					7
Marsiliaceæ.........	1			1				1
Musci............	48	134	6					140
Hepaticæ...........	15	16						16
Characeæ..........	2	10	1					11
Algæ.............	82	191	13					204
Lichens...........	30	131	15					146
Total.........	746	1472	129	200	8	245	7	2061

SUMMARY.

	FAMILIES.	GENERA	NATIVE.		NATURALIZED.		ADVENTIVE.		TOTAL.
			Species.	Varieties	Species.	Varieties	Species.	Varieties	
Exogens, ...	87	415	615	40	167	3	203	5	1033
Endogens, ...	20	125	324	46	32	5	42	2	451
Pteridophytes, . . .	5	20	51	8	1				60
Bryophytes, ...	2	63	150	6					156
Thallophytes, . . .	3	123	332	29					361
Total, . . .	117	746	1472	129	200	8	245	7	2061

INDEX.

Aaron's Rod, 36.
Abele, 96.
Abies, 96.
Abutilon, 18.
Acalypha, 90.
Acer, 22.
Achillea, 54.
Acnida, 86.
Acolium, 173.
Acorus, 98.
Actæa, 3.
Adam's Needle, 109.
Adder's Tongue, 137.
Adiantum, 135.
Adlumia, 5.
Æsculus, 22.
Æthusa, 41.
Agrimonia, 31.
Agrimony, 31.
Agropyrum, 133.
Agrostis, 127.
Ailanthus, 20.
Aira, 128.
Alder, 62, 64, 94.
Alectoria, 167.
Aletris, 105.
Alfalfa, 25.
Algæ, 152.
Alisma, 101.
Alismaceæ, 101.
Allium, 108.
Alnus, 94.
Alopecurus, 125.
Alsyke, 24.
Alyssum, 8.
Amarantaceæ, 85.
Amaranth, 85.
Amarantus, 85.

Amaryllis, 105.
Amaryllidaceæ, 105.
Ambrosia, 51.
Amelanchier, 34.
American Cowslip, 65.
Ammania, 39.
Ampelopsis, 21.
Amphicarpæa, 29.
Amsinckia, 75.
Anabæna, 163.
Anacardiaceæ, 21.
Anacharis, 102.
Anagallis, 66.
Anaphalis, 55.
Andromeda, 62.
Andropogon, 124.
Anemone, 1.
Anemonella, 1.
Angelica, 40.
Anomodon, 146.
Antennaria, 55.
Anthemis, 54.
Anthoxanthum, 125.
Anychia, 17.
Aphanorhegma, 143.
Aphanothece, 165.
Aphyllon, 67.
Apios, 29.
Apocynaceæ, 81.
Apocynum, 81.
Apple, 34.
Apple of Peru, 79.
Aquifoliaceæ, 64.
Aquilegia, 3.
Arabis, 6.
Araceæ, 97.
Aralia, 42.
Araliaceæ, 42.

Arbor-Vitæ, 97.
Arbutus, 61.
Archangelica, 40.
Arctium, 57.
Arctostaphylos, 61.
Arenaria, 15.
Arethusa, 103.
Argemone, 4.
Arisæma, 97.
Aristida, 125.
Aristolochiaceæ, 82.
Arrhenatherum, 128.
Arrow-grass, 101.
Arrow-head, 102.
Arrow-wood, 44.
Artemisia, 54.
Arthonia, 173.
Arthrodesmus, 161.
Arum, 97.
Asarum, 82.
Asclepiadaceæ, 81.
Asclepias, 81.
Ascophyllum, 154.
Ash, 82.
Asparagus, 108.
Aspen, 96.
Asperugo, 76.
Aspidium, 136.
Asplenium, 135.
Asprella, 134.
Aster, 47.
Atrichum, 144.
Atriplex, 84.
Aulacomnion, 144.
Avena, 128.
Avens, 31.
Azalea, 62.

Bachelor's Button, 56.
Bæomyces, 172.
Bæria, 53.
Ballota, 74.
Balm of Gilead, 96.
Balsam, 96.

Balsam Apple, 40.
Bambusina, 159.
Baneberry, 3.
Baptisia, 29.
Barbarea, 7.
Barberry, 3.
Barbula, 142.
Barley, 134.
Barnyard Grass, 123.
Bartonia, 80.
Bartramia, 143.
Basil, 71.
Basswood, 18.
Bastard Toad-flax, 89.
Batrachospermum, 153.
Bayberry, 93.
Bazzania, 150.
Beak-rush, 114.
Bearberry, 61.
Beard Grass, 124, 127.
Bedstraw, 44.
Beech, 93.
Beech-drops, 67.
Beech Fern, 136.
Beggar-ticks, 52.
Beggar's-lice, 76.
Beggiatoa, 164.
Bellis, 49.
Bellwort, 107.
Bengal Grass, 123.
Benjamin Bush, 89.
Bent Grass, 127.
Berberidaceæ, 3.
Berberis, 3.
Bergamot, 72.
Betonica, 73.
Betony, 69, 73.
Betula, 94.
Betulaceæ, 94.
Biatora, 172.
Bidens, 52.
Bindweed, 77, 87.
Birch, 94.
Bird Millet, 123.

Birthroot, 106.
Birthwort, 82.
Bitter Cress, 6.
Bittersweet, 78.
Blackberry, 32.
Black Grass, 109.
Black Horehound, 74.
Black Mustard, 8.
Black Snakeroot, 40.
Bladder Campion, 14.
Bladder Fern, 137.
Bladder Ketmia, 18.
Bladder Nut, 22.
Bladderwort, 66.
Bladderwrack, 154.
Blepharozia, 150.
Blephilia, 72.
Blite, 84.
Blitum, 84.
Bloodroot, 5.
Bloodwort, 105.
Blueberry, 61.
Bluebottle, 56.
Blue Curls, 70.
Blue Flag, 105.
Blue Grass, 130.
Blue Tangle, 61.
Bluets, 45.
Bœhmeria, 91.
Bog-rush, 109.
Boltonia, 49.
Boneset, 46.
Borage, 74.
Borraginaceæ, 74.
Borrago, 74.
Bottle-brush Grass, 134.
Bottle Grass, 123.
Bouncing Bet, 14.
Botrychium, 138.
Boxberry, 61.
Brachyelytrum, 126.
Brake, 135.
Brasenia, 4.
Brassica, 8.

Briza, 130.
Brizopyrum, 130.
Bromus, 132.
Brooklime, 68.
Brookweed, 66.
Broomrape, 67.
Bruchia, 140.
Brunella, 73.
Bryophytes, 139.
Bryopsis, 155.
Bryum, 143.
Buckbean, 81.
Buckthorn, 21.
Buckwheat, 86, 88.
Buellia, 172.
Bugleweed, 71.
Bugloss, 74.
Bulbochæte, 154.
Bulrush, 113.
Bunchberry, 42.
Bupleurum, 41.
Burdock, 57.
Bur Grass, 124.
Bur Marigold, 52, 53.
Burnet, 31.
Bur-reed, 98.
Bush Clover, 28.
Butter-and-Eggs, 67.
Buttercups, 2.
Butterflyweed, 81.
Butternut, 92.
Buttonbush, 45.
Buttonwood, 92.
Buxbaumia, 145.

Cactaceæ, 39.
Cactus, 39.
Cakile, 9.
Calamagrostis, 127.
Calamintha, 71.
Calicium, 173.
Calla, 97.
Callitrichaceæ, 90.
Callitriche, 90.

Calluna, 62.
Calopogon, 104.
Calothrix, 162.
Caltha, 3.
Calystegia, 78.
Camelina, 9.
Campanula, 60.
Campanulaceæ, 60.
Canary Grass, 124.
Cancer-root, 67.
Cannabis, 92.
Caprifoliaceæ, 43.
Capsella, 9.
Caraway, 42.
Cardamine, 6.
Cardinal Flower, 60.
Cardiospermum, 22.
Carex, 114.
Carpet-weed, 17.
Carpinus, 93.
Carrion Flower, 106.
Carrot, 40.
Carum, 42.
Carya, 92.
Caryophyllaceæ, 13.
Cashew, 21.
Cassandra, 62.
Cassia, 29.
Castanea, 93.
Castillea, 69.
Catchfly, 14.
Catnip, 72.
Cat-tail, 98.
Ceanothus, 21.
Cedar, 97.
Celandine, 5.
Celastraceæ, 22.
Celastrus, 22.
Celtis, 91.
Cenchrus, 124.
Centaurea, 56.
Cephalanthus, 45.
Ceramium, 153.
Cerastium, 16.

Ceratodon, 141.
Ceratophyllaceæ, 89.
Ceratophyllum, 89.
Cetraria, 166.
Chænactis, 53.
Chætophora, 156.
Chain Fern, 135.
Chamæcyparis, 97.
Chamomile, 54.
Chara, 151.
Characeæ, 151.
Charlock, 8.
Cheat, 132.
Checkerberry, 61.
Chelidonium, 5.
Chelone, 68.
Chenopodiaceæ, 83.
Chenopodium, 83.
Cherry, 30.
Chess, 132.
Chestnut, 93.
Chick-pea, 29.
Chickweed, 15, 16.
Chickweed Wintergreen, 65.
Chicory, 57.
Chimaphila, 63.
Chinquepin, 4.
Chiogenes, 61.
Chlorosporeæ, 155.
Chokeberry, 34.
Chondrus, 153.
Chorizanthe, 88.
Christmas Fern, 137.
Chroococcus, 165.
Chroolepus, 156.
Chrysanthemum, 54.
Chrysopogon, 124.
Chrysosplenium, 35.
Cichorium, 57.
Cicuta, 41.
Cinna, 127.
Cinnamon Fern, 137.
Cinquefoil, 31.
Circæa, 37.

Cirsium, 56.
Cistaceæ, 12.
Cladium, 114.
Cladonia, 171.
Cladophora, 156.
Clarkia, 38.
Clathrocystis, 165.
Claytonia, 17.
Clearweed, 91.
Cleavers, 44.
Clematis, 1.
Clethra, 62.
Climacium, 146.
Climbing Bittersweet, 22.
Climbing Fern, 137.
Climbing Hempweed, 46.
Clintonia, 107.
Closterium, 162.
Clotbur, 51.
Clover, 23.
Club Moss, 138.
Club-rush, 113.
Cnicus, 56.
Cocklebur, 51.
Cœlosphærium, 165.
Cohosh, 3.
Colic-root, 105.
Coliseum Ivy, 67.
Collema, 169.
Collinsonia, 72.
Coltsfoot, 46.
Columbine, 3.
Comandra, 89.
Comfrey, 74.
Commelynaceæ, 110.
Compass-plant, 51.
Compositæ, 45.
Comptonia, 94.
Cone Flower, 52.
Conferva, 157.
Coniferæ, 96.
Conium, 42.
Conjugatæ, 159.
Conotrema, 171.

Convallaria, 107.
Convolvulaceæ, 77.
Convolvulus, 77.
Coptis, 3.
Corallorhiza, 104.
Cord Grass, 124.
Coreopsis, 52.
Cornaceæ, 42.
Corn Cockle, 15.
Cornel, 42.
Cornus, 42.
Coronilla, 29.
Corydalis, 5.
Corylus, 93.
Coscinodon, 142.
Cosmarium, 160.
Costmary, 54.
Cotton-grass, 113.
Couch Grass, 133.
Cowbane, 41.
Cowberry, 61.
Cow-herb, 14.
Cow-lily, 4.
Cow-parsnip, 40.
Cow-wheat, 70.
Crab Grass, 121.
Cranberry, 61.
Cranberry Tree, 44.
Cranesbill, 19.
Crassulaceæ, 35.
Cratægus, 33.
Crepis, 58.
Cress, 5.
Crotalaria, 23.
Crowfoot, 1, 2.
Cruciferæ, 5.
Cryptogamia, 135.
Cryptotænia, 41.
Cucumber-root, 107.
Cucurbitaceæ, 40.
Cudweed, 55.
Cupressus, 97.
Cupuliferæ, 92.
Currant, 35.

Cuscuta, 78.
Cut Grass, 124.
Cylindrospermum, 165.
Cylindrothecium, 146.
Cynoglossum, 76.
Cyperaceæ, 111.
Cyperus, 111, 175.
Cypripedium, 104.
Cystopteris, 137.

Dactylis, 130.
Daisy, 49.
Dalibarda, 32.
Dandelion, 59.
Dangleberry, 61.
Danthonia, 128.
Daphne, 89.
Darnel, 133.
Datura, 79.
Daucus, 40.
Day Lily, 109.
Dead-nettle, 73.
Delesseria, 153.
Delphinium, 3.
Dentaria, 6.
Deschampsia, 128.
Desmidieæ, 159.
Desmidium, 159.
Desmodium, 27.
Dewberry, 32.
Deyeuxia, 127.
Dianthus, 13.
Dicentra, 5.
Dichelyma, 145.
Dicksonia, 137.
Dicranella, 140.
Dicranum, 140.
Diervilla, 43.
Dioscorea, 106.
Dioscoreaceæ, 106.
Diphyscium, 145.
Diplachne, 129.
Diplopappus, 48.
Dirca, 89.

Distichlys, 130.
Ditch-grass, 99.
Docidium, 161.
Dock, 88.
Dodder, 78.
Dodecatheon, 65.
Dogbane, 81.
Dog's-tail Grass, 128.
Dog's-tooth Violet, 108.
Dogwood, 21, 42.
Draba, 8.
Dragon-head, 73.
Draparnaldia, 156.
Drop-seed Grass, 126.
Drosera, 12.
Droseraceæ, 12.
Drummondia, 142.
Duckweed, 98.
Dulichium, 112.
Dulse, 153.
Dutchman's Breeches, 5.
Dwarf Dandelion, 57.
Dyer's Weed, 23.

Eatonia, 129.
Echinacea, 51.
Echinocystis, 40,
Echinodorus, 101.
Echinospermum, 75.
Echium, 74.
Ectocarpus, 155.
Eel-grass, 99, 102.
Elatinaceæ, 13.
Elatine, 13.
Elder, 44.
Elecampane, 51.
Eleocharis, 112.
Eleusine, 128.
Elm, 91.
Elodea, 13.
Elodes, 13.
Elymus, 134.
Enchanter's Nightshade, 37
Endocarpon, 173.

Endogens, 97.
Ephemerum, 140.
Epigæa, 61.
Epilobium, 37.
Epiphegus, 67.
Equisetaceæ, 135.
Equisetum, 135.
Eragrostis, 129.
Erechtites, 55.
Ericaceæ, 60.
Erigeron, 48.
Eriocaulon, 111.
Eriocaulonaceæ, 111.
Eriophorum, 113.
Eritrichium, 75.
Erodium, 19.
Erysimum, 7.
Erythronium, 108.
Euastrum, 160.
Eupatorium, 46.
Euphorbia, 90.
Euphorbiaceæ, 90.
Evening Primrose, 37.
Everlasting, 55.
Evernia, 166.
Exogens, 1.

Fagopyrum, 88.
Fagus, 93.
Fall Dandelion, 58.
False Flax, 9.
False Spikenard, 107.
Featherfoil, 66.
Ferns, 135.
Fescue, 131.
Festuca, 131.
Feverfew, 54.
Ficoideæ, 17.
Figwort, 67, 68.
Filices, 135.
Fimbriaria, 149.
Fimbristylis, 114.
Finger Grass, 121.
Fir Balsam, 96.

Fireweed, 55.
Fissidens, 141.
Five-finger, 31.
Flag 98, 105.
Flax, 19.
Fleabane, 49.
Floating Heart, 81.
Florideæ, 152.
Flowering Dogwood, 42.
Flowering Fern, 137.
Fontinalis, 145.
Fool's Parsley, 41.
Forget-me-not, 75.
Four-o'clock, 82.
Fowl Meadow Grass, 130.
Foxglove, 69.
Foxtail, 123, 125.
Fragaria, 32.
Fraxinus, 82.
Frog's-bit, 102.
Frostweed, 12.
Frullania, 150.
Fucus, 154.
Fuirena, 112.
Fumaria, 5.
Fumariaceæ, 5.
Fumitory, 5.
Funaria, 143.

Galeopsis, 73.
Galingale, 111.
Galinsoga, 53.
Galium, 44.
Gall-of-the-earth, 59.
Garget, 83.
Garlic, 108.
Gastridium, 127.
Gaultheria, 61.
Gaylussacia, 60.
Gelidium, 152.
Genista, 23.
Gentian, 80.
Gentiana, 80.
Gentianaceæ, 80.

Geraniaceæ, 19.
Geranium, 19.
Gerardia, 69.
Germander, 70.
Geum, 31.
Gilia, 77.
Ginger, 82.
Ginseng, 42.
Glaux, 66.
Glœocapsa, 165.
Glœocystis, 158.
Glœothece, 165.
Gloiotrichia, 162.
Glyceria, 130.
Gnaphalium, 55.
Goat's Rue, 27.
Golden Ragwort, 56.
Goldenrod, 50.
Goldthread, 3.
Goodyera, 103.
Gooseberry, 34.
Goosefoot, 83.
Goose-grass, 44.
Goose-tongue, 54.
Gourd, 40.
Gracilaria, 153.
Gramineæ, 121.
Grape, 21.
Graphis, 173.
Grass, 121.
Grass of Parnassus, 35.
Gratiola, 68.
Green Briar, 106.
Grimaldia, 149.
Grimmia, 142.
Grindelia, 50.
Gromwell, 75.
Ground Hemlock, 97.
Ground Ivy, 73.
Ground Nut, 20, 42.
Ground Pine, 138.
Groundsel, 55.
Gum-plant, 50.
Gymnostichum, 134.

Gypsophila, 14.

Habenaria, 102.
Hackberry, 91.
Hackmatac, 96.
Hæmodoraceæ, 105.
Hair Grass, 127, 128.
Halorageæ, 36.
Hamamelaceæ, 36.
Hamamelis, 36.
Hardhack, 31.
Harebell, 60.
Hawkweed, 58.
Hawthorn, 33.
Hazlenut, 93.
Heal-all, 73.
Heart's ease, 11.
Heath, 60, 62.
Hedeoma, 71.
Hedgehog Grass, 124.
Hedge Hyssop, 68.
Hedge Mustard, 7.
Hedwigia, 142.
Helenium, 53.
Helianthemum, 12.
Helianthus, 52.
Heliophytum, 76.
Heliotrope, 76.
Heliotropium, 76.
Hellebore, 107.
Hemerocallis, 109.
Hemizonia, 53.
Hemlock, 96, 97.
Hemp, 81, 92.
Hemp-nettle, 73.
Hemp-weed, 46.
Henbane, 79.
Hepatica, 1.
Hepaticæ, 149.
Heracleum, 40.
Herb Robert, 19.
Herd's Grass, 126.
Hesperis, 10.
Hibiscus, 18.

Hickory, 92.
Hieracium, 58.
Hierochloa, 125.
Hildenbrantia, 153.
Hobble-bush, 44.
Hog-peanut, 29.
Holcus, 128.
Holly, 64.
Honewort, 41.
Honeysuckle, 43.
Hop, 92.
Hop-tree, 20.
Hordeum, 133.
Horehound, 73, 74.
Hornbeam, 93.
Hornwort, 89.
Horse Chestnut, 22.
Horse-Gentian, 43.
Horse Nettle, 78.
Horseradish, 6.
Horsetail, 135.
Horseweed, 48.
Hottonia, 66.
Hound's Tongue, 76.
Houseleek, 36.
Houstonia, 45.
Huckleberry, 60.
Humulus, 92.
Hungarian Grass, 123.
Hyalotheca, 159.
Hydrocharidaceæ, 102.
Hydrocotyle, 40.
Hydrophyllaceæ, 76.
Hydrophyllum, 76.
Hyoscyamus, 79.
Hypericaceæ, 13.
Hypericum, 13.
Hypnum, 146.
Hypochæris, 58.
Hypoxys, 105.

Ilex, 64.
Ilicineæ, 64.
Ilysanthes, 68.

Impatiens, 20.
India Wheat, 88.
Indian Grass, 124.
Indian Hemp, 81.
Indian Pipe, 63.
Indian Poke, 107.
Indian Rice, 124.
Indian Tobacco, 60.
Indian Turnip, 97.
Innocence, 45.
Inula, 51.
Ipomœa, 77.
Iridaceæ, 105.
Iris, 105.
Irish Moss, 153.
Ironweed, 45.
Ironwood, 93.
Isatis, 9.
Isoeteæ, 139.
Isoetes, 139.
Isopyrum, 2.
Iva, 51.
Ivy, 21, 67, 73.

Jack-in-the-pulpit, 97.
Jerusalem Artichoke, 52.
Jerusalem Oak, 83.
Jewel-weed, 20.
Jointed Charlock, 10.
Joint Grass, 127.
Jointweed, 87.
Juglandaceæ, 92.
Juglans, 92.
Juncaceæ, 109.
Juncus, 109.
Juneberry, 34.
Juniper, 97.
Juniperus, 97.

Kale, 8.
Kalmia, 62.
Knapweed, 56.
Knawel, 17.
Knotgrass, 87.

Krigia, 57.

Labiatæ, 70.
Labrador Tea, 63.
Lactuca, 59.
Ladies' Tobacco, 55.
Ladies' Tresses, 103.
Lady Fern, 136.
Lady's Slipper, 104.
Lady's Thumb, 86.
Lambkill, 62.
Lamium, 73.
Lampsana, 57.
Laportea, 91.
Lappa, 57.
Larch, 96.
Larix, 96.
Larkspur, 3.
Lathyrus, 28.
Lauraceæ, 89.
Laurel, 62, 89.
Lavender, 65.
Layia, 53.
Leadwort, 65.
Leatherleaf, 62.
Leatherwood, 89.
Leathesia, 155.
Lecanora, 170.
Lechea, 12.
Lecidea, 172.
Ledum, 63.
Leek, 108.
Leersia, 124.
Leguminosæ, 23.
Lemna, 98.
Lemnaceæ 98.
Lentibulariaceæ, 66.
Leontodon, 58.
Leonurus, 73.
Lepidium, 9.
Lepigonum, 16.
Leptobryum, 143.
Leptochloa, 128.
Leptodon, 145.

Leptogium, 169.
Leptothrix, 164.
Leptotrichum, 141.
Lespedeza, 28.
Lettuce, 59.
Leucanthemum, 54.
Leucobryum, 141.
Leucodon, 146.
Leucothoe, 61.
Leverwood, 93.
Liatris, 45.
Lichens, 165.
Ligusticum, 41.
Ligustrum, 82.
Lilac, 82.
Liliaceæ, 106.
Lilium, 108.
Lily, 106, 108.
Lily of the Valley, 107.
Limnanthemum, 81.
Linaceæ, 19.
Linaria, 67.
Linden, 18.
Lindera, 89.
Linnæa, 43.
Linum, 19.
Lion's-foot, 59.
Liparis, 104.
Liquorice, 45.
Lithospermum, 75.
Live-forever, 36.
Liverwort, 149.
Lobelia, 60.
Lobeliaceæ, 60.
Locust, 27.
Lolium, 133.
Lonicera, 43.
Loosestrife, 39, 65.
Lophanthus, 72.
Lophocolea, 150.
Lopseed, 70.
Lousewort, 69.
Lovage, 41.
Lucerne, 25.

Ludwigia, 38.
Lungwort, 75.
Lupine, 23.
Lupinus, 23.
Luzula, 109.
Lychnis, 15.
Lycium, 79.
Lycopersicum, 78.
Lycopodiaceæ, 138.
Lycopodium, 138.
Lycopsis, 74.
Lycopus, 71.
Lygodium, 137.
Lyme Grass, 134.
Lyngbya, 163.
Lysimachia, 65.
Lythraceæ, 39.
Lythrum, 39.

Madder, 44.
Madotheca, 150.
Maianthemum, 108.
Maidenhair, 135.
Mallow, 17.
Malva, 17.
Malvaceæ, 17.
Mandrake, 3.
Maple, 22.
Marchantia, 149.
Marigold, 52, 53.
Marrubium, 73.
Marsh Bell-flower, 60.
Marsh Cress, 6.
Marsh Marigold, 3.
Marsh Rosemary, 65.
Marsilia, 139.
Marsiliaceæ, 139.
Maruta, 54.
Mastigonema, 162.
Matricaria, 54.
Matrimony Vine, 79.
May Apple, 3.
Mayflower, 61.
Mayweed, 54.

Meadow Beauty, 39.
Meadow Grass, 130.
Meadow Parsnip, 41.
Meadow Rue, 1.
Meadow-sweet, 30.
Medeola, 107.
Medicago, 25.
Medick, 25.
Melampyrum, 70.
Melastoma, 39.
Melastomaceæ, 39.
Melilot, 25.
Melilotus, 24.
Mentha, 70.
Menyanthes, 81.
Mercury, 21, 90.
Mermaid-weed, 37.
Mertensia, 75.
Metzgeria, 150.
Mexican Tea, 83.
Mezereum, 89.
Micrasterias, 160.
Microcoleus, 163.
Microseris, 58.
Microstylis, 104.
Mikania, 46.
Milkweed, 81.
Milkwort, 22, 66.
Millet, 123.
Mimulus, 68.
Mint, 70.
Mint Geranium, 54.
Mitchella, 45.
Mitella, 35.
Mithridate Mustard, 9.
Mitrewort, 35.
Mnium, 144.
Mockernut, 92.
Mollugo, 17.
Monarda, 72.
Moneses, 63.
Moneywort, 65.
Monkey-flower, 68.
Monostroma, 157.

Monotropa, 63.
Moonwort, 138.
Moosewood, 89.
Morning-glory, 77.
Morus, 91.
Mosses, 139.
Mossy Stonecrop, 36.
Motherwort, 73.
Mountain Ash, 34.
Mountain Rice, 126.
Mugwort, 54.
Muhlenbergia, 126.
Mulberry, 91.
Mulgedium, 59.
Mullein, 67.
Musci, 139.
Mustard, 5, 7, 8, 9.
Myosotis, 75.
Myriangium, 171.
Myrica, 93.
Myricaceæ, 93.
Myriophyllum, 36.

Nabalus, 59.
Naiadaceæ, 99.
Naias, 99.
Nardia, 150.
Nasturtium, 5.
Neckera, 145.
Neckweed, 60.
Neillia, 30.
Nelumbium, 4.
Nemopanthes, 64.
Nepeta, 72.
Nephroma, 169.
Nesæa, 39.
Nettle, 91.
New Jersey Tea, 21.
Nicandra, 79.
Nicotiana, 80.
Nightshade, 78.
Nine-bark, 30.
Nipplewort, 57.
Nitella, 151.

Nodularia, 163.
Nonesuch, 25.
Nostoc, 163.
Nuphar, 4.
Nut-rush, 114.
Nyctaginaceæ, 82.
Nymphæa, 4.
Nymphæaceæ, 4.
Nyssa, 43.

Oak, 92.
Oakesia, 107.
Oat, 128.
Oat Grass, 126, 128.
Œnothera, 37, 175.
Old-maid's frizzles, 31.
Old-witch Grass, 121.
Oleaceæ, 82.
Olive, 82.
Onagraceæ, 37.
Onoclea, 137.
Onopordon, 57.
Oosporeæ, 154.
Opegrapha, 173.
Ophiocytium, 158.
Ophioglossum, 137.
Opuntia, 39.
Orchard Grass, 130.
Orchidaceæ, 102.
Orchis, 102.
Ornithogalum, 108.
Orobanchaceæ, 67.
Orpine, 35.
Orthocarpus, 69.
Orthotrichum, 142.
Oryzopsis, 126.
Oscillaria, 164.
Osier, 95.
Osmorrhiza, 42.
Osmunda, 137.
Ostrich Fern, 137.
Ostrya, 93.
Oswego Tea, 72.
Oxalis, 20.
Ox-eye Daisy, 54.

Oxybaphus, 82.

Painted-cup, 69.
Panic Grass, 121.
Panicum, 121.
Pannaria, 169.
Papaveraceæ, 4.
Pansy, 11.
Parietaria, 92.
Parmelia, 167.
Parnassia, 35.
Paronychiæ, 17.
Parsley, 40, 41.
Parsnip, 40.
Parthenium, 51.
Partridge-berry, 45.
Partridge-pea, 29.
Paspalum, 121.
Pastinaca, 40.
Pear, 34.
Pearlwort, 16.
Pediastrum, 158.
Pedicularis, 69.
Pellia, 149.
Pellitory, 92.
Peltandra, 97.
Peltigera, 169.
Penium, 162.
Pennyroyal, 71.
Pennywort, 40.
Penthorum, 35.
Pentstemon, 68.
Pepperbush, 62.
Peppergrass, 9.
Pepperidge, 43.
Peppermint, 70.
Pepper-root, 6.
Periploca, 82.
Pertusaria, 170.
Petalostemon, 26.
Petunia, 79.
Phacelia, 76.
Phænogamia, 1.
Phæosaccion, 155.

Phæosporeæ, 154.
Phalaris, 124.
Phegopteris, 136.
Philonotis, 143.
Phleum, 126.
Phlox, 77.
Phragmites, 129.
Phryma, 70.
Phyllitis, 154.
Physalis, 79.
Physcia, 168.
Physcomitrella, 140.
Physcomitrium, 143.
Physma, 169.
Physostegia, 73.
Phytolacca, 83.
Phytolaccaceæ, 83.
Picea, 96.
Pickerel-weed, 110.
Pigeon Grass, 123.
Pigweed, 83, 85.
Pilea, 91.
Pimpernel, 66, 68.
Pine, 96.
Pine-sap, 63.
Pink, 13.
Pinus, 96.
Pinweed, 12.
Pinxter-flower, 63.
Pipewort, 111.
Pipsissewa, 63.
Pirus, 34.
Pisum, 29.
Pitcher-plant, 4.
Placodium, 169.
Plane-tree, 92.
Plantaginaceæ, 64.
Plantago, 64.
Plantain, 64.
Platanaceæ, 92.
Platanus, 92.
Pleurocarpus, 159.
Pleurotænium, 161.
Pluchea, 51.

Plum, 30.
Plumbaginaceæ, 65.
Poa, 130.
Podophyllum, 3.
Podostemaceæ, 90.
Podostemon, 90.
Pogonatum, 145.
Pogonia, 104.
Poison Hemlock, 42.
Poison Ivy, 21.
Poke, 83.
Polemoniaceæ, 77.
Polygala, 22.
Polygalaceæ, 22.
Polygonaceæ, 86.
Polygonatum, 108.
Polygonum, 86.
Polypodium, 135.
Polypody, 135.
Polypogon, 127.
Polysiphonia, 152.
Polytrichum, 145.
Pond Lily, 4.
Pondweed, 99.
Pontederia, 110.
Pontederiaceæ, 110.
Poplar, 96.
Poppy, 4.
Populus, 96.
Porphyra, 153.
Porphyridium, 158.
Portulaca, 17.
Portulacaceæ, 17.
Potamogeton, 99.
Potato, 78.
Potentilla, 31.
Poterium, 31.
Pottia, 141.
Poverty Grass, 125.
Prairie Clover, 26.
Prairie Dock, 51.
Prenanthes, 59.
Prickly Ash, 20.
Prickly-pear, 39.

Primrose, 65.
Primulaceæ, 65
Prince's Feather, 86.
Prince's Pine, 63.
Privet, 82.
Proserpinaca, 37.
Protococcaceæ, 158.
Protococcus, 158.
Prunus, 30.
Ptelea, 20.
Pteridophytes, 135.
Pterigynandrum, 146.
Pteris, 135.
Pulse, 23.
Punctaria, 154.
Purple Cone-flower, 51.
Purslane, 17.
Pycnanthemum, 71.
Pylaisia, 146.
Pyrenula, 174.
Pyrola, 63.
Pyxine, 168.

Quaking Grass, 130.
Queen of the Meadow, 46.
Quercus, 92.
Quick Grass, 133.
Quillwort, 139.
Quitch Grass, 133.

Racomitrium, 142.
Radish, 10.
Radula, 150.
Ragweed, 51.
Ralfsia, 155.
Ramalina, 166.
Ranunculaceæ, 1.
Ranunculus, 2.
Raphanus, 10.
Raspberry, 32.
Rattle-box, 28.
Rattlesnake Grass, 130.
Rattlesnake Plantain, 103.
Rattlesnake-weed, 58.

Ray Grass, 133.
Red-top, 130.
Reed, 129.
Reed Grass, 127.
Rhamnaceæ, 21.
Rhamnus, 21.
Rhexia, 39.
Rhizoclonium, 156.
Rhododendron, 62.
Rhodora, 63.
Rhodymenia, 153.
Rhus, 21.
Rhynchospora, 114.
Ribbon Grass, 125.
Ribes, 34.
Ribgrass, 64.
Riccia, 149.
Richweed, 72, 91.
Rinodina, 170.
Riverweed, 90.
Rivularia, 162.
Robinia, 27.
Robin's Plantain, 49.
Rock Cress, 6, 7.
Rocket, 7, 9, 10.
Rock-rose, 12.
Roman Wormwood, 51.
Rosa, 33.
Rosaceæ, 30.
Rose, 33.
Rose Acacia, 27.
Rose Mallow, 18.
Rosin-weed, 51.
Roxbury Waxwork, 22.
Royal Fern, 137.
Rubiaceæ, 44.
Rubus, 32.
Rudbeckia, 52.
Rue, 20.
Rumex, 88.
Ruppia, 99.
Rush, 109.
Rutaceæ, 20.
Rye, 133, 134.

Rye Grass, 133.

Sabbatia, 80.
Sacheria, 153.
Sagina, 16.
Sagittaria, 102.
Salicaceæ, 94.
Salicornia, 84.
Salix, 94.
Salsola, 85.
Salt Grass, 124.
Salt-marsh Fleabane, 51.
Salt-marsh Grass, 124.
Saltwort, 85.
Salvia, 72.
Sambucus, 44.
Samolus, 66.
Samphire, 84.
Sandalwood, 89.
Sand Grass, 129.
Sandwort, 15.
Sanguinaria, 5.
Sanicula, 40.
Santalaceæ, 89.
Sapindaceæ, 22.
Saponaria, 14.
Sarracenia, 4.
Sarraceniaceæ, 4.
Sarsaparilla, 42.
Sassafras, 89.
Satureia, 71.
Savin, 97.
Saxifraga, 35.
Saxifragaceæ, 34.
Saxifrage, 35.
Scheuchzeria, 101.
Scapania, 150.
Scarlet-fruited Thorn, 34.
Scilla, 108.
Scirpus, 113.
Scleranthus, 17.
Scleria, 114.
Scouring-rush, 135.
Scrophularia, 68.

Scrophulariaceæ, 67.
Scorpiurus, 29.
Scutellaria, 73.
Scytonema, 162.
Scytosiphon, 155.
Sea Milkwort, 66.
Sea Rocket, 9.
Secale, 133.
Sedge, 114.
Sedum, 36.
Seed-box, 38.
Selaginella, 138.
Self-heal, 73.
Sempervivum, 36.
Senebiera, 9.
Seneca Grass, 125.
Senecio, 55.
Sensitive Fern, 137.
Sericocarpus, 47.
Setaria, 123.
Shadbush, 34.
Shagbark Hickory, 92.
Shave Grass, 135.
Sheep-berry, 44.
Shepherd's Purse, 9.
Shield Fern, 136.
Shin-leaf, 63.
Shooting-star, 65.
Sickle-pod, 7.
Sicyos, 40.
Sida, 18.
Side-saddle flower, 4.
Silene, 14.
Silphium, 51.
Silver-weed, 31.
Simarubaceæ, 20.
Sisymbrium, 7.
Sisyrinchium, 105.
Sium, 41.
Skullcap, 73.
Skunk cabbage, 98.
Smartweed, 87.
Smilacina, 107.
Smilaceæ, 106.

Smilax, 106.
Snake-head, 68.
Snakeroot, 40, 46.
Sneeze-weed, 53.
Snowberry, 61.
Soapberry, 22.
Soapwort, 14.
Solanaceæ, 78.
Solanum, 78.
Solidago, 49.
Solomon's Seal, 108.
Sonchus, 59.
Sorghum, 124.
Sorrel, 88.
Sow-thistle, 59.
Spanish Needles, 53.
Sparganium, 98.
Spartina, 124.
Spear Grass, 130, 131.
Spearmint, 70.
Spearwort, 2.
Specularia, 60.
Speedwell, 68.
Speirodela, 98.
Spergula, 16.
Spergularia, 16.
Sphærozosma, 159.
Sphærozyga, 163.
Sphagnaceæ, 149.
Sphagnum, 149.
Spice-bush, 89.
Spiderwort, 110.
Spike Grass, 130.
Spikenard, 42, 107.
Spike-rush, 112.
Spiræa, 30.
Spiranthes, 103.
Spirogyra, 159.
Spirulina, 164.
Spleenwort, 135.
Spondylosium, 159.
Sporobolus, 126.
Spring Beauty, 17.
Spring Cress, 6.

Spruce, 96.
Spurge, 90.
Spurrey, 16.
Squirrel-tail Grass, 133.
Stachys, 73.
Staff Tree, 22.
Staphylea, 22.
Star-cucumber, 40.
Star-flower, 65.
Star Grass, 105.
Star-of-Bethlehem, 108.
Starwort, 90.
Statice, 65.
Staurastrum, 161.
Steetzia, 149.
Steironema, 65.
Stellaria, 15.
Stereocaulon, 171.
Stitchwort, 15.
Stickseed, 75.
Sticta, 168.
Stigeoclonium, 156.
Stipa, 126.
St. John's-wort, 13.
Stonecrop, 35, 36.
Stone-root, 72.
Strawberry, 32.
Streptopus, 107.
Struthiopteris, 137.
Suæda, 85.
Succory, 57.
Sugarberry, 91.
Sumach, 21.
Summer Savory, 71.
Sundew, 12.
Sunflower, 52.
Swamp Honeysuckle, 62.
Swamp Pink, 62.
Sweet-brier, 33.
Sweet Cicely, 42.
Sweet Clover, 25.
Sweet-fern, 94.
Sweet Flag, 98.
Sweet Gale, 93.

Sweet William, 14.
Sycamore, 92.
Symphytum, 74.
Symplocarpus, 98.
Syringa, 82.

Tamarack, 96.
Tanacetum, 54.
Tansy, 54.
Tape-grass, 102.
Taraxacum, 59.
Tare, 28.
Taxus, 97.
Tear-thumb, 87.
Tephrosia, 27.
Tetmemorus, 161.
Tetraphis, 143.
Tetraspora, 158.
Teucrium, 70.
Thalictrum, 1.
Thallophytes, 151.
Thaspium, 41.
Thelia, 146.
Theloschistes, 167.
Thimbleberry, 32.
Thin Grass, 127.
Thistle, 56.
Thlaspi, 9.
Thorn-apple, 79.
Thoroughwort, 46.
Three-seeded Mercury, 90.
Thuya, 97.
Thyme, 71.
Thymeleaceæ, 89.
Thymus, 71.
Tiarella, 35.
Tickseed Sunflower, 52.
Tidy-tips, 53.
Tilia, 18.
Tiliaceæ, 18.
Timothy, 123, 126.
Toad-flax, 67, 89.
Tolypothrix, 162.
Tomato, 78.

Tower Mustard, 7.
Tradescantia, 110.
Trapa, 37.
Tree of Heaven, 20.
Trichostema, 70.
Tricuspis, 129.
Trientalis, 65.
Trifolium, 23.
Triglochin, 101.
Trigonella, 26.
Trillium, 106.
Triosteum, 43.
Triplasis, 129.
Triticum, 133.
Tropidocarpum, 10.
Trumpet-weed, 46.
Tsuga, 96.
Tupelo, 43.
Tussilago, 46.
Twayblade, 104.
Twig-rush, 114.
Twin-flower, 43.
Typha, 98.
Typhaceæ, 98.

Ulmus, 91.
Ulota, 142.
Ulothrix. 157.
Ulva, 157.
Umbelliferæ, 40.
Umbilicaria, 168.
Umbrella-grass, 113.
Urceolaria, 171.
Urtica, 91.
Urticaceæ, 91.
Usnea, 166.
Utricularia, 66.
Uvularia, 107.

Vaccaria, 14.
Vaccinium, 61.
Vallisneria, 102.
Vanilla Grass, 125.
Vaucheria, 154.

Velvet Grass, 128.
Velvet-leaf, 18.
Venus' Looking-glass, 60.
Veratrum, 107.
Verbascum, 67.
Verbena, 70.
Verbenaceæ, 70.
Vernal Grass, 125.
Vernonia, 45.
Veronica. 68.
Verrucaria, 173.
Vervain, 70.
Vetch, 28.
Vetchling, 28.
Viburnum, 44.
Vicia, 28.
Vilfa, 126.
Vincetoxicum, 82.
Vine, 21.
Viola, 10, 175.
Violaceæ, 10.
Violet, 10, 175.
Virginia Creeper, 21.
Virginian Cowslip, 75.
Virgin's Bower, 1.
Vitaceæ, 21.
Vitis, 21.
Volvox, 158.

Wake Robin, 106.
Wall Barley, 134.
Walnut, 92.
Water Chestnut, 37.
Water Chinquepin, 4.
Water Cress, 5.
Water-fern, 139.
Water Hemlock, 41.
Waterleaf, 76.
Water Lily, 4.
Water Marigold, 53.
Water Milfoil, 36.
Water Oats, 124.
Water Parsnip, 41.
Water Pepper, 87.

Water Plantain, 101.
Water Purslane, 38.
Water Shield, 4.
Water Starwort, 90.
Waterweed, 102.
Waterwort, 13.
Wax Myrtle, 93.
Way Bent, 134.
Webera, 143.
Weisia, 140.
Wheat, 133.
Wheat Grass, 133.
White Alder, 62.
White Grass, 124.
White Hellebore, 107.
White Mustard, 8.
White Snakeroot, 46.
Whiteweed, 54.
Whitewood, 18.
Whitlow Grass, 8.
Whitlow-wort, 17.
Wicopy, 89.
Wild Bean, 29.
Wild Indigo, 29.
Wild Oats, 107.
Wild Pink, 14.
Wild Senna, 29.
Wild Sensitive-plant, 29.
Willow, 94.
Willow-herb, 37.
Wind-flower, 1.
Winterberry, 64.
Winter Cress, 7.
Wintergreen, 61, 63, 65.
Wire Grass, 128, 130.
Witch Hazel, 36.
Withe-rod, 44.
Woad, 9.

Woad-waxen, 23.
Wood Betony, 73.
Woodbine, 21.
Wood Fern, 136.
Wood Grass, 124.
Wood-rush, 109.
Wood Sage, 70.
Wood Sorrel, 20.
Woodsia, 137.
Woodwardia, 135.
Wool-grass, 113.
Worm-seed Mustard, 7.
Wormwood, 55.

Xanthidium, 161.
Xanthium, 51.
Xanthorrhiza, 3.
Xanthoxylum, 20.
Xylographa, 173.
Xyridaceæ, 110.
Xyris, 110.

Yam, 106.
Yarrow, 54.
Yellow Adder's-tongue, 108.
Yellow Cress, 6.
Yellow-eyed-grass, 110.
Yellow Mustard, 8.
Yellow Nelumbo, 4.
Yellow-root, 3.
Yew, 97.
Yucca, 109.

Zannichellia, 99.
Zizania, 124.
Zizia, 41.
Zostera, 99.
Zygnema, 159.